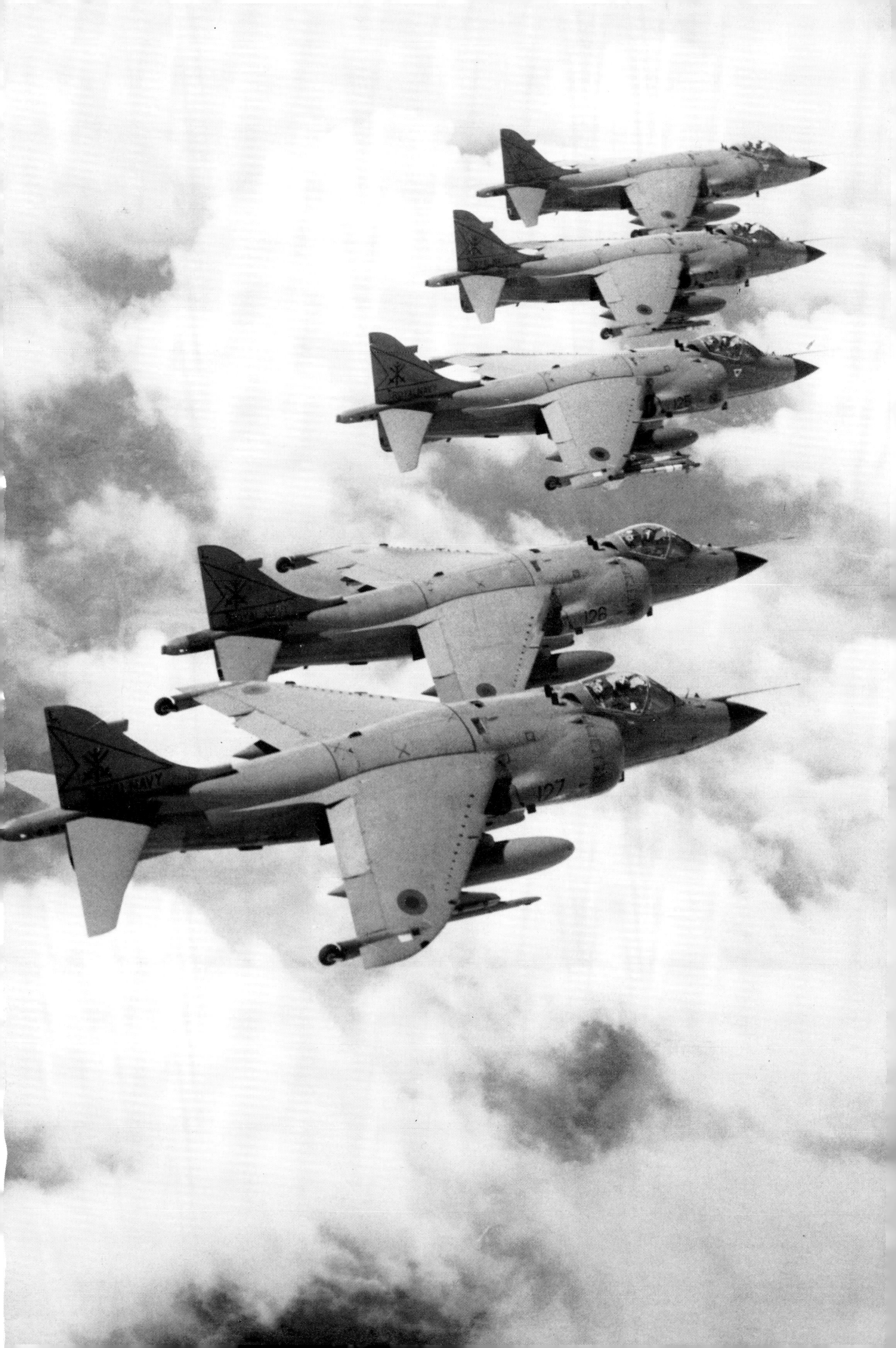
126
127

BRITISH AEROSPACE
A PROUD HERITAGE

Geoff Green

First Published 1988

ISBN 0 9510519 2 X

All rights reserved. No part of this book may be reproduced or transmitted in any form or by any means, electronic or mechanical, including photo-copying, recording or by any information storage and retrieval system, without permission from the publisher in writing.

© Geoff Green 1988

Published by
Geoff Green,
37 Underhill Road,
Charfield,
Wotton-under-Edge,
Glos. GL12 8TQ

Photoset and printed by
Printing and Graphic Services (Bristol) Ltd.
Golf Course Lane,
Filton,
Bristol BS12 7QS

Frontispiece:
Royal Navy Sea Harriers.

Title Page:
G-AHRF, the prototype Vickers Viscount 630 airliner was a world beater.

Contents

Introduction	5
Avro	6
Bristol	16
De Havilland	26
Blackburn	36
Vickers	46
Hawker	56
Folland	66
English Electric	74
Chester	82
Scottish Aviation	88
Absorbed Names	96
British Aircraft Corporation	98
Hawker Siddeley	106
British Aerospace	114

Acknowledgements

My name appears on the front cover but many people have helped in the preparation of this book. Firstly, I must thank Mel West of Printing & Graphic Services for supporting the idea of a BAe history.

Working for British Aerospace has given me direct access to the public relations offices throughout the Company. For supplying photographs and source material as well as checking typescripts and tolerating numerous enquiries I am grateful to Dr. Norman Barfield, Eric Barker, Howard Berry, Terry Bickerton, Phillip Birtles, Mike Brown, David Dorman, John Ellis, Geoff Hill, Harry Holmes, John Humby, Karen Roden, Wally Rouse, Mike Stroud and Penny Telling.

For the supply or loan of photographs and source material I wish to thank other BAe personnel including Vince Balaam, Richard Bridgland, Dave Charlton, Darryl Cott, Elaine Eddolls, Hugh Field, Sean Greene, Gerry Hammond, Nigel Oldfield, Karen Mitchell, Sue Turner, Pete Vincent and Alan Yates.

All photographs are by courtesy of British Aerospace unless credited otherwise.

Two people contributed the results of their own private research. I am grateful to Tony Sedgwick for providing the manuscript from which the Hamble text was taken and to Alan Robertson for allowing me access to his work on Scottish Aviation; full details of which will be found in his book entitled ''Lion Rampant and Winged''.

My engineering contacts at the various sites have been leaned on for help and local information and here I should mention Jim Dyson, Gerry Fry, Colin Greenslade, John Jones, Dennis Lloyd, Fred Maginn, George Mercer, Vic Mould, Harry Punter, Tony Rose, Garth Sylvester and Willy Taggart.

Rolls Royce Bristol Engine Division came to the rescue on numerous occasions and I wish to thank Jack Barnard, John Hutchinson and David Howie.

Proof reading is most important in ensuring the accuracy of the content and I have been fortunate to have had specialist help from John Heaven and Dave Oyns.

If I have inadvertently missed anyone out, thank you very much.

Geoff Green

Preface

The purpose of this book is to commemorate British Aerospace's first decade as the premier U.K. aerospace manufacturer and to try to give a brief insight into the next ten years as the long term corporate plans reach maturity.

In describing the lineage up to 1960, the theme has been to pick out the most successful aircraft (a top twenty in essence) of each constituent company in terms of sales volume or importance to aviation progress in general. Then to bring the story up to date by recording the achievements of British Aircraft Corporation, Hawker Siddeley and British Aerospace.

Writing a history such as this is fraught with difficulty because the evolution of British aircraft is a complex subject; it becomes a case not of what to put into it but what to leave out. Available space precludes an individual history of every famous company, but so that they do not go unmentioned I have included a couple of views of aircraft by some of the most important independent firms.

This is not an official history and any opinions expressed herein are my own and not those of BAe or any other company.

Introduction

It is over eighty years since A.V. Roe took his first faltering steps in British aviation and it was to take several years more before flyers could go wherever and whenever they pleased almost regardless of the overpowering effect of wind direction.

Balloons had been a source of curiosity and experiment in Britain since 1784 but without positive means of controlling height and direction their use was limited and operation hazardous.

Heavier than air machines attracted the attention of several learned men in Britain in the nineteenth century. One of the earliest experimenters was Sir George Cayley, a Yorkshire squire who, like Henson, Wenham and Stringfellow after him, was concerned with large gliders. Sir Hiram Maxim, a U.S. Citizen living in England did much to further the understanding of the principles of flight; he registered patents and built several machines but never succeeded in making one fly. The first flight of an untethered man-carrying glider is reputed to have taken place at Cardross, Dumbartonshire in 1895.

As soon as heavier than air flight had become a reality many companies, some large and others only 'one man bands' sprang up like mushrooms after rain. Not all have survived the test of time and some newcomers now extant joined the race at half distance.

Under market and governmental pressure they have been merged or eliminated one by one. There were about 80 aircraft firms in existence after the First World War but by 1945 this number had been reduced to 27 airframe builders and 7 sources of aero-engines in the U.K.

By the middle of the 1960s 'shotgun marriages' had reduced this still further to two major airframe builders and one engine maker.

Now, over three quarters of a century after the first flights, the number of firms engaged in both aspects of the business has been reduced to unity.

Technology too has advanced. In 1910 a single seat aeroplane flying one hundred feet above the ground was an achievement; today it is an everyday occurrence for airliners to fly five miles high carrying hundreds of passengers. Speeds have increased from a running pace to three times the speed of sound and continue to rise.

The first fifty years was dominated by fixed wing aeroplanes but, since discovery of the great strides made by Germany in the last war, rockets have replaced aircraft for many military purposes. Helicopters have been developed for special duties and space vehicles have left the pages of science fiction books and become a reality.

British Aerospace is made up of the nine foremost aviation companies in Great Britain which via an interim existence as British Aircraft Corporation or Hawker Siddeley now make up the manufacturing strength of BAe producing a wide range of aerospace products.

This is their story; there are others, which for commercial or political reasons exist on the fringes, but these can only be mentioned briefly in passing.

In the following pages will be found the aircraft and a few ballistic vehicles, past, present and future that represent British Aerospace's proud heritage.

AVRO

... Manchester

The name AVRO, an acronym of A.V. Roe, became famous for aircraft in two world wars and in the jet age. However, the man who founded it did not control the destiny of the company of that name in later years and it is with others that the credit must be shared.

Alliott Verdon Roe, son of a Manchester doctor was born in Patricroft in 1877. Alert yet modest, he was impetuous as a boy having only a moderate liking for school and sought adventure.

In 1892 at the age of 15 he travelled to Canada to seek his fortune surveying in British Columbia but found only unemployment there. Returning to Britain a year later he became apprenticed to the Lancashire & Yorkshire Railway at its Locomotive Works in Horwich. Later still, having studied marine engineering at King's College, London he went to sea as an engineer with the South African Mail Company.

He is quoted as saying that he first became interested in flight by watching Albatrosses during quieter moments at sea and made model gliders in 1898 which he launched off the side of the ship.

Leaving the sea in 1902 he took employment as a draughtsman with the motor car firm of Brotherton & Crockers Limited but continued to explore the theory of flight in his spare time.

Concerned about the lack of progress with heavier than air machines in Britain, A.V. Roe wrote a letter to The Times pointing out the lead held by the Wright Brothers of America who had flown at Kittyhawk, North Carolina on 17 December 1903. His letter was printed in the Times Engineering Supplement on 21 January 1906 together with a rebuff that 'artificial aviation' was 'dangerous to human life' and 'doomed to failure'.

In that year he became Secretary to the Aero Club but soon resigned and travelled to the USA to work on the Davidson Gyrocopter financed by Sir W.G. Armstrong Whitworth & Company. His stay in the USA was short also.

In 1906 Lord Northcliffe offered the first of several prizes for aeroplanes to promote sales of his 'Daily Mail' newspaper. A.V., as he was known to his friends, submitted three model designs and secured a prize of £75.

Spurred on by this success he proceeded to translate his models into man-carrying size at his brother, Dr. S. Verdon Roe's house in Putney. At this juncture the owners of Brooklands race track, seeking publicity, offered a prize of £2,500 for the first aircraft to fly round the oval track before the end of 1907. A.V. built all the parts of a biplane himself and erected it in a shed at Brooklands. Underpowered by the 9hp JAP engine it would not fly but with a larger Antoinette engine Roe achieved a series of hops at Brooklands on 8 June 1908. After six 'flights' the Brooklands owners grew tired of aviation and Roe was evicted; his precious biplane was thrown over a fence to clear the track.

Undeterred by the Philistines and his own slow progress, two new machines, triplanes this time, were started in his brother's stables but needing more elbow room A.V. searched extensively and eventually set up shop under the arches of the Great Eastern Railway at Lea Marshes in Essex. One triplane was sold incomplete but the other was 'flown' on 13 July 1909; this and subsequent efforts gaining him the nickname of "Roe the Hopper". It was later painted yellow and named *Bulls Eye* to advertise the elastic braces made by the company run by his brother Humphrey. A.V. had obtained financial help from Humphrey and a partnership was formed on 27 April 1909.

Evicted once again, this time by the local council A.V. moved to Wembley Park for further flying attempts.

Alliott realised his ambition to become an aircraft manufacturer when A.V. Roe & Company was registered in January 1910. Construction was undertaken in a basement of Brownfield Mills,

Above:
A.V. ROE
Alliott Verdon Roe (later to receive a Knighthood) photographed in 1909 in front of the Avroplane, his first triplane design.

Left:
AVRO VULCAN
Prototype 'straight wing' Vulcan in 1950s.

Above:
ROE TRIPLANE
A.V. Roe favoured the triplane for his early attempts at manned flight. Nine different types were built and one of the two Roe II machines is shown here.

Manchester run by Everards in which Humphrey Verdon Roe had a large interest.

Seeing aviation in a new light and as a source of revenue the owners of Brooklands race track developed the grassed area enclosed by the concrete track itself as a flying ground. Brooklands was to become the Mecca of sporting aviation up to the start of the Great War. Hangars were erected for rental and No. 14 was acquired by A.V.Roe & Company for its flying base.

A Roe *Mercury* triplane was exhibited at the Aero & Motor Boat Show held at Olympia, London in March 1910 but later crashed at Brooklands.

By the middle of 1910 the *Roe IV* biplane, actually a triplane with truncated lower wings, was flown at Brooklands and was soon followed by a true biplane known as the Type D. A development of this, Type E, was flown on 12 April 1912.

On 11 January 1913, A.V.Roe & Company Limited was formed with additional capital including that from a local brewing family. Extra workshop space was obtained in Clifton Street, Miles Platting.

At about this time a 19-year-old youth, Roy Chadwick, joined A.V.Roe as a draughtsman. His technical skill and personal integrity was to be of great value to the Company in future years. His first task was redesigning the Type E which then became the Avro 500 and marked the start of a new era in Avro history. It flew on 8 May 1912.

The Avro Type F, first flown on 1 May 1912, was Avro's first monoplane and the world's first aircraft to have a fully enclosed cockpit. It shared some features with the Avro 500 and was built for a Portuguese customer.

At the start of the 1914-18 War larger premises were rented from Mather & Platt Limited at Newton Heath, Manchester (Park Works) for the mass production of the Avro 504. This machine was a resounding success with well over eight thousand being produced: a just reward for Alliott Verdon Roe's faith in aviation in the difficult opening years of the twentieth century. Over four thousand Avro 504's were built in the Avro factories in Manchester and

Below:
AVRO 504
G-EADL, an Avro 504K, (formerly registered E4348) was built by Eastbourne Aviation Co. Ltd. It became a flying test bed for the Cosmos (later Bristol) Lucifer 3-cylinder engine.

Above:
AVRO 555 BISON
The second prototype Bison, serial N154, with two-bay wings and Napier Lion engine as built in 1922. 56 Bisons were built as Fleet gunnery spotters and torpedo layers.

Hamble; the remainder being built by no less than twenty-four subcontractors in Britain, Canada, Denmark, Sweden and Japan.

Subsequent military types were aimed at turning the 504 into a fighter but none got past the prototype stage.

Avro's first twin engined aircraft, appropriately named *Manchester*, was flown in December 1918 but only two were completed.

When Roy Chadwick assumed the post of Chief Designer in 1919, he and his team faced the difficult task of preparing ideas for peacetime aircraft. With large numbers of aircraft including Avro 504's, Bristol F2B's and Sopwith biplanes readily available at knock down prices from the Disposals Board the manufacturing prospect was not good by any means.

Chadwick put his faith in a small lightweight private owner biplane called *Baby* which was first flown, and crashed, on 30 April 1919. Nine were built altogether in single and two seat form known variously as *Baby*, *Water Baby*, *Racer* and *Antarctic Baby* to mention a few.

To take advantage of surplus airframes the Type 536 five passenger tourer was a 504 conversion. G-EACX (formerly K134) was the first of twenty-two to this design.

In order to supplement the meagre amount of available aircraft work Avro arranged a deal with Crossley Motors, another Manchester firm, with a view to getting motor vehicle work but this was not successful.

A part of S.E. Saunders Limited at Cowes, famous at the time for its 'Consuta' brand of marine plywood sewn with copper wire, was bought by Alliott in 1921 from Vickers. The latter had obtained a controlling interest in the Isle of Wight firm in 1918. Renamed Saunders Roe Limited it gave A.V. the opportunity to satisfy his urge to build seaplanes. From this date, his interest in Avro diminished and he spent the rest of his working life promoting Saunders Roe.

After being in the doldrums since the Armistice, aircraft manufacturing began to pick up generally from 1922 onward. A variety of aeroplanes was created at Avro in the next five years or so.

Chadwick's *Aldershot* was a large biplane powered by one Rolls Royce *Condor* engine and was built to an Air Ministry specification for a long range day bomber in competition with the D.H.27 *Derby*. Two prototypes were built at Avro's Hamble factory and were followed in modified form by fifteen production *Aldershot Mk.III's* to equip No.99 squadron at Andover in 1924.

In competition with the Blackburn *Blackburn,* the Avro 555 *Bison* fleet spotter and reconnaissance biplane first flew in 1921 and sales to the Fleet Air Arm were a welcome boost to Avro's finances. The *Aldershot* and *Bison* were to be the lifeblood of Avro between 1922 and 1927.

Avro test flying in the Manchester area was done at Alexandra Park but in 1924 New Hall Farm, at Woodford, Cheshire was purchased. Hangars were erected there together with a grass aerodrome which has been expanded over the years and is still used for Manchester built aircraft.

Having achieved success with a bomber for the RAF and a spotter for the Navy, Chadwick pursued the idea of a fighter for the RAF. Built as a private venture, the *Avenger* was flown by Bert Hinkler in June 1926 but the front line fighter was to elude Manchester to the end. Like the *Buffalo* torpedo bomber the *Avenger*

Above:
G-EBVZ, a production Avian III with 85hp Cirrus II engine. 195 Avians were built plus 205 of the de Havilland Gipsy engined type 616 Avian IVs. Successful flights were made to Australia and South Africa and many Avians were exported.

Right:
AUTOGIRO
In the mid-1920s, Avro built a series of experimental autogiros for the Air Ministry to test the ideas of Don Juan de la Cierva.

Below:
AVRO 626
J301 and J302 with Armstrong Siddeley Lynx engines in Egyptian Air Force markings. A development of the Tutor for general purpose duties, almost 200 Avro 626's were built including a trainer version known as the Prefect.

was to remain a one off and joined the collection of Avro 'might have beens' of the 1920s.

Turning once again to the light civil market, Chadwick devised the *Avian* for the 1926 Daily Mail Light Aeroplane Competition and after exchanging its original Armstrong Siddeley *Genet* engine for an ADC *Cirrus* it attracted good sales.

In 1928 Alliott Verdon Roe finally severed his ties with A.V.Roe & Company Limited when he sold his share of the business to John Siddeley whose Armstrong Siddeley Development Company thus gained control of Avro.

Antony Fokker in Holland had gained worldwide acclaim for his high wing monoplanes and, eager to break into the large civil market, Avro purchased a licence to build the Fokker.F.VII. This resulted in the Avro *Ten* of which fourteen were built between 1928 and 1930.

The smaller Avro 619 *Five* designed by Chadwick as a five passenger version of the *Ten* had A.S.*Genet* engines. Five were built plus two of a six passenger version, Type 624. The experience gained in building large monoplanes was to set Avro on the right road for the future; but Manchester had not yet finished with biplanes.

Always mindful of the success of the 504 series as a trainer, Chadwick set out on paper his ideas for a successor to it in 1929. Having a fabric covered tubular steel airframe and powered by a 5-cylinder A.S.*Mongoose* engine the two seat Type 621 *Trainer* entered service with the RAF in 1930. Later deliveries of the Avro 621 with 7-cylinder *Lynx* engines were known as *Tutors* and were built until 1936.

The Avro 626 was a fighter version of the *Tutor* for smaller air forces for which duty it was armed with fore and aft firing Vickers guns and could carry bombs. Some Type 626 *Prefect* trainers were also built.

When the Armstrong Siddeley Development Company was merged with part of Hawker Aircraft Limited to form Hawker Siddeley Aircraft Company Limited on 11 July 1935, Avro became a constituent of what was to become Britain's most influential aircraft and engineering group prior to 1977.

Avro then bade farewell to biplanes and, starting with the Type 642 *Eighteen,* concentrated on monoplane transports. Of the two that were built the first one, *G-ACFV,* had two A.S.*Jaguar* engines and was furnished for eighteen passengers on inter-city services. The other Type 642 had four A.S.*Lynx* engines and was supplied as a VIP aircraft to the Viceroy of India.

Two Type 652 low wing monoplanes derived from the *Eighteen* and its Fokker predecessors were built for Imperial Airways in 1934 as four-passenger couriers. They were joined later by a 652 Mk.II supplied to the Egyptian Government in 1936.

The 652 seemed set to be just another low volume aircraft but its adoption as the *Anson* by the RAF in 1935 for coastal patrol and training duties started a production run that was to surpass that of the Avro 504 during the next seventeen years.

Armstrong Siddeley *Cheetah* engines were the standard power plant but *Anson II's* and *Anson V's,* both built in Canada, had Jacobs and Pratt & Whitney engines respectively.

Below:
AVRO ANSON
The last of 10,994 Ansons to be built pictured flying near Woodford aerodrome in 1952. WJ561 was an Anson T21 navigation trainer.

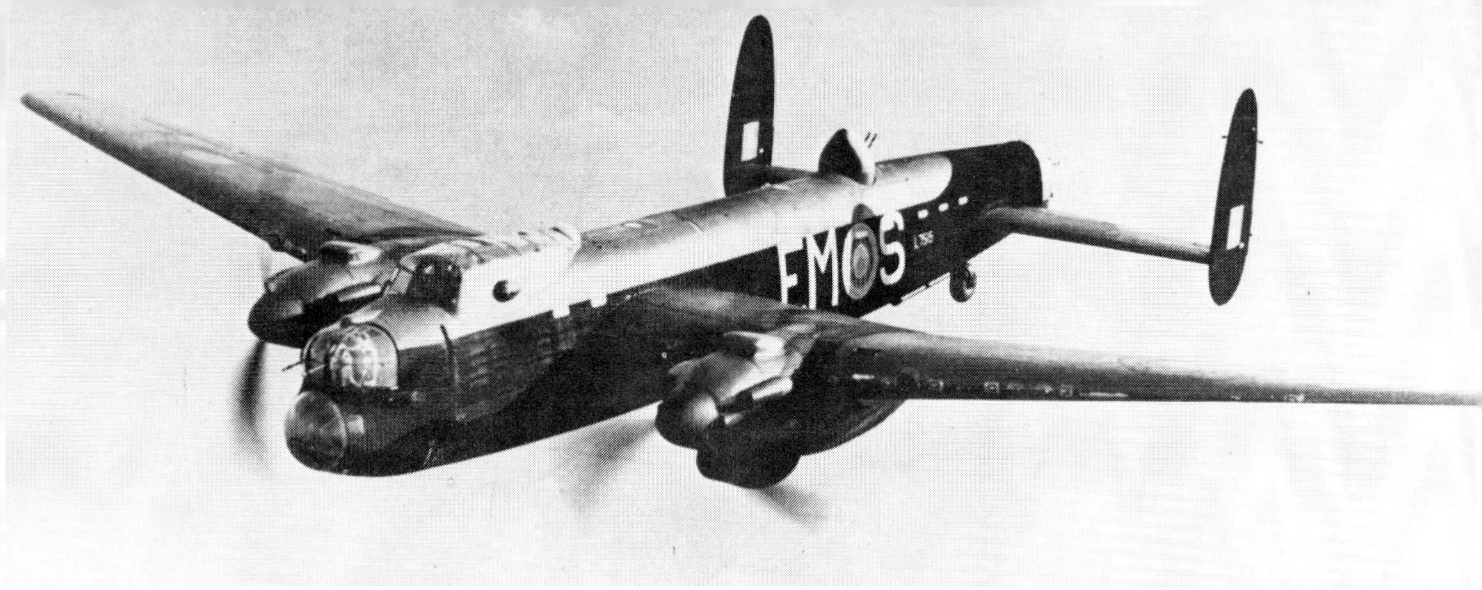
FM S

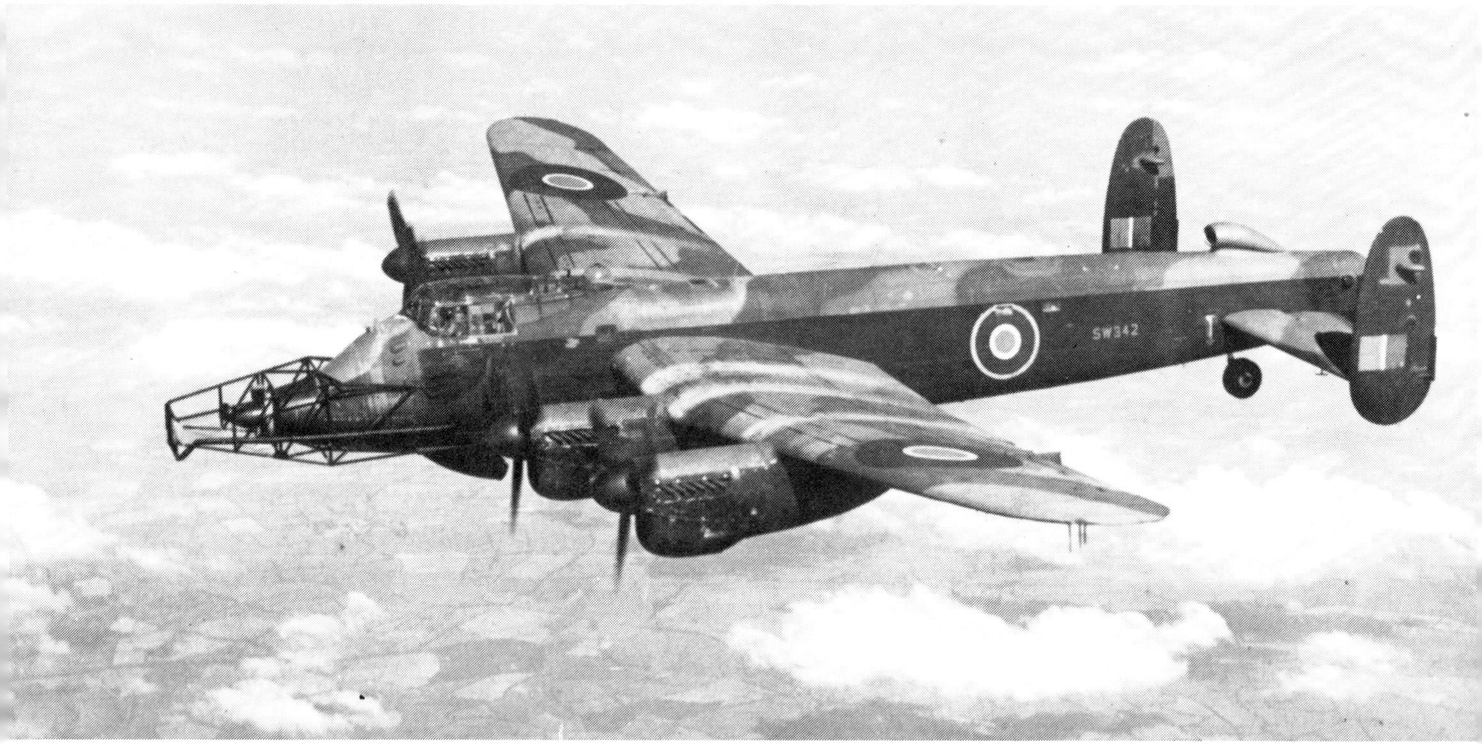
SW342

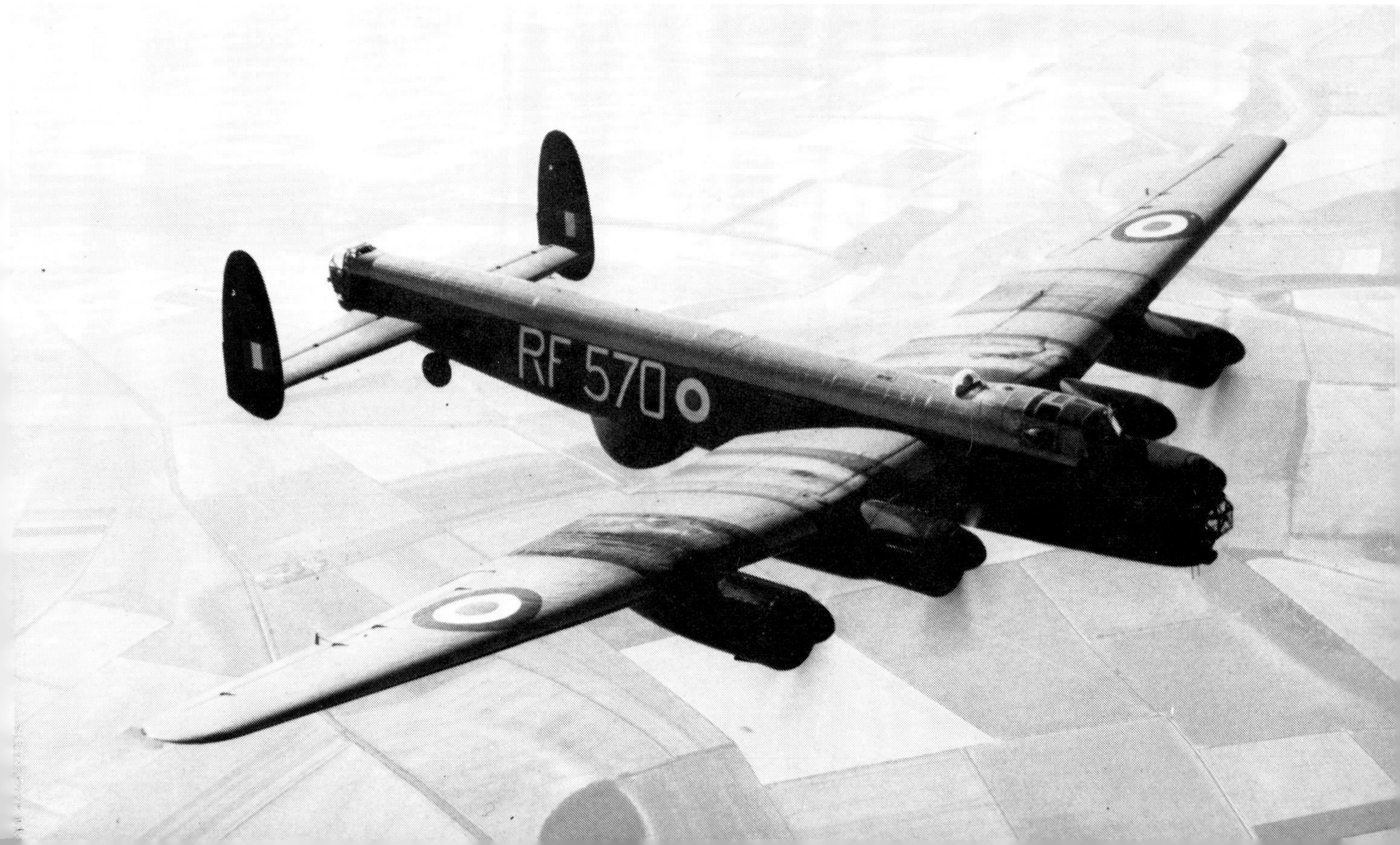
RF 570

As *Anson* production got underway, Chadwick and his team embarked on an ambitious project that was to evolve into the best heavy bomber of the Second World War.

The prototype *Manchester*, *L7246,* flew on 25 July 1939 and went into production in modified form in 1940. *Manchesters* first witnessed the heat of battle in 1941 but problems with its two Rolls Royce *Vulture* engines caused a major redesign.

Now powered by four Rolls Royce *Merlins* and renamed *Lancaster* it flew as a prototype on 9 January 1941. As the RAF implemented 'Bomber' Harris' plans for "thousand bomber raids" on German industrial targets *Lancasters* replaced *Wellingtons* as quickly as they could be built.

The *Lancaster's* ability to carry very heavy loads such as Barnes Wallis' earthquake bombs and Dam Buster's spinning bomb together with the skill and bravery of RAF Bomber Command aircrews made it a legend in its time.

To cope with wartime production Chadderton, which had earlier produced Blenheims, and the Newton Heath Works were augmented by new dispersal factories. In the 1930s the Ivy Works was established in an old five storey mill at Failsworth between Oldham and Manchester. Each floor had its own trade, the top one being a woodmill, and was used in the latter part of its occupation for *Ansons.*

For *Lancasters,* a shadow factory and airfield at Yeadon, near Leeds came under Avro control . The Empire Works, also at Failsworth, was used for sheet metal details.

Ansons, Blenheims, Manchesters and *Lancasters* were all taken by road to Woodford for final assembly and pre delivery flight tests.

Originally known as the *Lancaster B.IV* and *B.V.* the *Lincoln* was the definitive bomber variant intended for long range operation against the Japanese. *PW925,* the prototype flew on 9 June 1944 and *Lincolns* went into RAF service in September 1945 by which time the war was over.

After the defeat of Germany and Japan, *Lincolns* formed the backbone of Bomber Command until the arrival of the 'V' bomber force in the mid-1950s. They were very useful for proving the new turboprop and turbojet engines and were assigned to engine manufacturers as flying test beds or to special RAF squadrons for 'jet' familiarisation.

Having been responsible one way or another for the production of over seven thousand Lancasters, Avro was well qualified to build large multi-engined transports and this formed a sound basis for post war civil development.

Conversion of the *Lancaster* was a natural progression and on the eve of the ending of the war in Europe, *Lancasters* in progress came off the line as the *Lancastrian* mail 'plane and as thirteen passenger courier transports for the RAF. Long range *Lancastrians* were operated by British South American Airways.

By combining the wings and tail of the *Lancaster* with a new slab sided fuselage the *York* was born. The wings were set higher than on the *Lancaster* to allow a row of circular windows below the mainplane to afford passengers a better downward view. First of the *Yorks*, *LV626*, was flown in January 1944, one year before the *Lancastrian*, and they were sold to BOAC and other carriers requiring quick delivery.

Left Upper:
AVRO MANCHESTER
L7515 was a Manchester 1A with twin tail fins. Due to problems with its Rolls Royce Vulture engines only 157 Manchesters were built and the remainder of this batch (43 aircraft) were built as a new type, the Lancaster.

Left Centre:
AVRO LANCASTER
A unique machine; SW342, a Yeadon built Lancaster B.III as a flying test bed with no less than six engines. Besides the usual Rolls Royce Merlins it had an Armstrong Siddeley Mamba turboprop in the nose and an A.S. Adder in the tail. *Rolls Royce.*

Left Lower:
AVRO LINCOLN
RF570, a Lincoln B.II with Rolls Royce Merlin engines was built by Armstrong Whitworth at Coventry.

Below:
AVRO YORK
Merlin engined York C.1 in RAF service. Derived from the Lancaster, the York gave valuable airline service in the late 1940s until the new post war airliners became available. 257 Yorks were built including one in Canada.

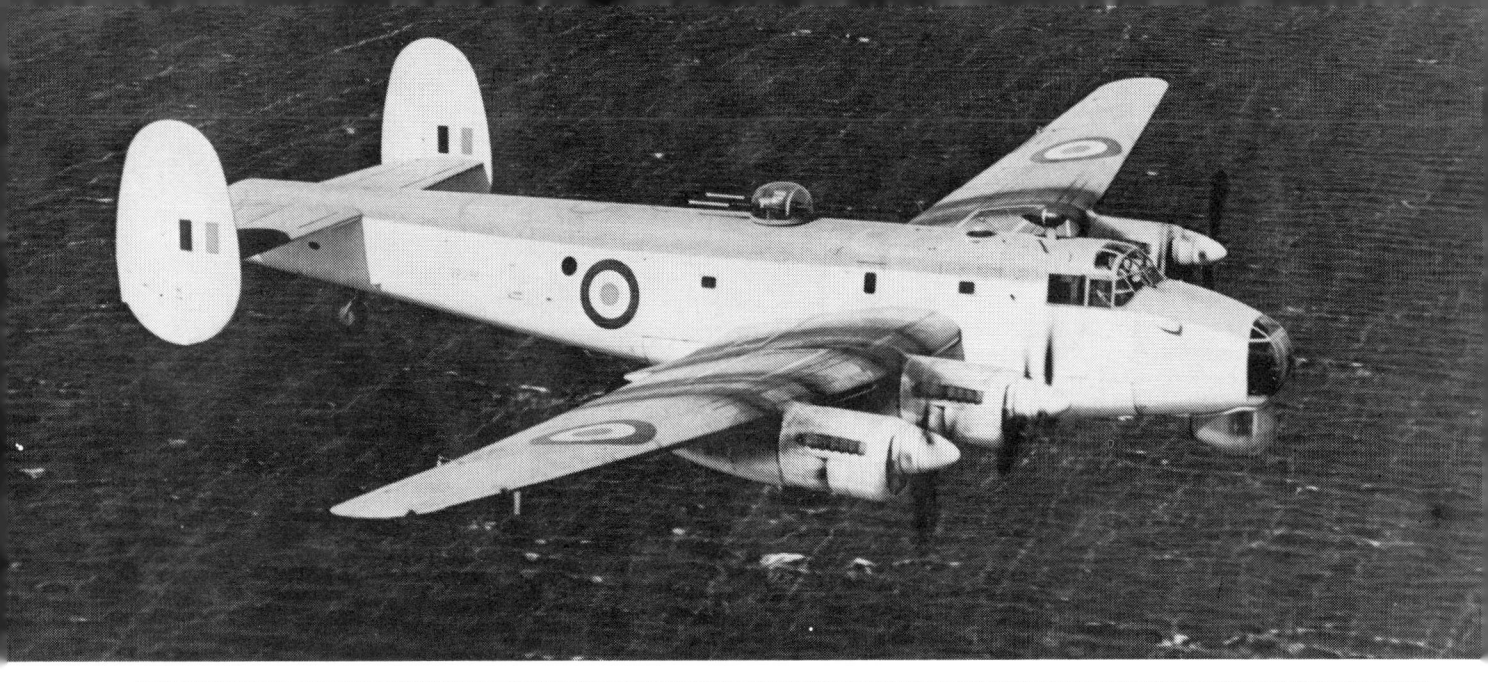

Top:
AVRO SHACKLETON
   VP258 from the initial production batch of 29 Shackleton M.R.I's (Maritime Reconnaissance) on a test flight. Shackletons entered RAF service with Coastal Command in 1951.

Above:
AVRO TUDOR
   Merlin engined Tudor 2, G-AGSU, was first flown with Ministry of Supply serial TT185 on 10 March 1946. Thirty-five Tudors were built in eight basic versions including the turbojet engined Tudor 8.

Below:
AVRO ASHTON
   WB492 was one of six Ashtons built for the Ministry of Supply in 1949 for experimental work.

The pressurised airliner pioneered in the USA was a new breed of machine with a new set of problems to solve. A contemporary of the Vickers *Viking*, Avro's interim type the *Tudor* with its circular fuselage was a marked departure from past efforts.

Sales volume of the *Tudor* was low partly due to engineering problems and partly as a result of procrastination and specification changes by BOAC. Nevertheless, the *Tudor* was Britain's first passenger machine to have a pressurised cabin and gave experience which was to prove invaluable ten years later.

Conversion of the second *Tudor I* prototype to have Rolls Royce *Nene* turbojets in a pair of siamesed pods beneath the wings produced Avro's first jet aircraft, the *Tudor 8*. VX195 flew from Woodford on 6 September 1948.

The success of the *Tudor 8* as a development tool prompted the Ministry of Supply to order six more. Part finished *Tudor* airframes were completed in a similar configuration to the *Tudor 8* and named *Ashton*. They had an active career in radar and instrument testing. An *Ashton 3* (WB493) was a familiar sight at Filton in the 1950s for high altitude research of the Bristol *Olympus* engines being developed for the *Vulcan*; it even became a 'star' in the film "Cone of Silence."

The *Vulcan* was to be Avro's contribution to the nuclear bomber deterrent force and sustained the Company through the 'fifties and early 'sixties; it was intended to follow this with a supersonic version but Government cost cutting put paid to that ambition.

To support the *Vulcan* programme five *707* experimental aircraft were built. The first of these one-third scale delta machines, VX794, first flew on 4 September 1950 to research low-speed handling. It was joined by another low speed *707* on the following day. The high speed *707A* flew in June 1951. One *707* was destroyed in a fatal crash but the other four were used extensively in the 1950s and three of them are now preserved.

While the *Vulcan* was designed to fly high and fast for hit and run nuclear bombing missions to the Soviet Union, the *Shackleton* was a low and slow maritime reconnaissance aircraft capable of staying on patrol for many hours. First flown on 9 March 1949 it was a descendant of the *Lincoln*.

When the supersonic bomber project was cancelled in 1957 as a result of the Defence White Paper, Manchester was building the larger *Vulcan B.2*'s but did not have a project to succeed it. Taking over more or less where the *Tudor* left off, the Avro 748 with two Rolls Royce *Dart* turboprops was first envisaged as a short or medium range airliner to carry fifteen to twenty passengers. The prototype flew in June 1960.

Already being a subsidiary of the Hawker Siddeley Group, the Manchester firm was spared from the Government pressure that was being applied to its competitors. With production of the *Vulcan B.2* in full swing and the *748* showing good promise for the future the people at Chadderton and Woodford barely noticed the change from Avro to becoming a member of Hawker Siddeley Aviation.

Above:
AVRO VULCAN
XL321, one of the larger Vulcan B.2's armed with a replica of the rocket powered Avro Blue Steel nuclear stand-off bomb.

Below:
AVRO 748
The second prototype 748, G-ARAY, in company with G-APZV the first prototype. G-APZV which first flew on 24 June 1960 was converted to the prototype 748MF Andover (G-ARRV) while G-ARAY had an active life as a demonstrator and was later converted to Series 2.

"Bristol"
... Focus On Filton

"Bristol"

The British & Colonial Aeroplane Company Limited was founded in Bristol on 19 February 1910 and a factory was set up in a bus depot at Filton, Gloucestershire, about a mile outside the northern boundary of the city of Bristol.

Sir George White, a Bristol born entrepreneur and millionaire had been responsible for the tram networks in various parts of the country including Bristol, London, Dublin, Coventry and Middlesborough. He foresaw a profitable future for the sport of aviation which was then in its infancy. As France was the centre of activity of this new invention a licence was taken out to build *Zodiac* biplanes at Filton. A complete aircraft was bought from France and it was intended to build a pilot batch of five more. The *Zodiac* was taken to Brooklands but all attempts to make it fly resulted in failure.

George Challenger, works manager at Filton and son of Charles Challenger the General Manager of Bristol Tramways built a biplane based on the Farman 'boxkite' from the parts of the *Zodiac*. Bristol *Biplane No.7* made its maiden flight at Larkhill on Salisbury Plain on 30 July 1910.

Production quickly got underway and by the end of 1910 ten *Biplanes* had been built including an *Extended Type* with extensions to the upper wing tips. Two of these were shipped to India and demonstrated early in 1911 and the *Biplane* was also flown in Australia. Besides the *Biplanes* built for the Bristol Flying Schools and the War Office, sales were also made to the Russian Government.

Not being an engineer or an aviator himself, Sir George recruited the best people he could find to design aeroplanes to build in his new factory. Frenchman Pierre Prier came to Filton in June 1911 and in the next two years was responsible for a series of successful single and two seat monoplanes similar to Bleriot's machines.

Other aeroplanes were designed by Grandseigne, Archibald Low, Gordon-England and Commander Burney but by far the greatest contribution prior to the First World War was that of Henri Coanda. The son of the Roumanian war minister, Coanda came to Bristol in 1912 after working with Eiffel in Paris. He continued the work of Pierre Prier initially but then proceeded to introduce his own innovations. When monoplanes met with official disfavour he turned to biplanes instead.

Bristol's marketing strategy was very effective. Sir George was staunchly patriotic and had contact with the British Royal Family. Most of the countries in Europe were ruled or influenced by the descendants of Queen Victoria and the armies of these countries were easily persuaded to buy *Bristol* aeroplanes. Sales were made to Russia, Germany, Spain, Italy and Roumania. Pattern machines were supplied to Deutsche Bristol Werke, Caproni in Italy and Breguet in France for local manufacture.

The pre-war years had seen many changes of faces but in 1914 Frank Barnwell began to exert himself as a designer in his own right. With the help of L.G.Frise and later Wilfred Reid and Clifford Tinson he established a permanent design office at Filton. These men were to be responsible for practically all of the aeroplanes built at Filton in the subsequent thirty years.

Capt. Barnwell's first accredited design was the *Baby Tractor Biplane No.206* which was developed into a series of *Scout* biplanes that gave valiant service in the Great War.

Several experimental types were also built but the most famous machine of the British & Colonial era was undoubtedly the Bristol *Fighter* popularly known as the "Brisfit". By the time that peace had returned to Europe no less than five thousand aeroplanes had been built at Filton in addition to Bristol types built by subcontractors. Over the next forty years Bristol was only just able to match that total.

The British & Colonial Aeroplane Company Limited went into voluntary liquidation early in 1920 and was reconstituted on 9 February 1920 as The Bristol Aeroplane Company Limited with a nominal capital of one million pounds.

With a surplus of aircraft the 'twenties were lean years for aircraft manufacturers and in an effort to remain viable Filton built 'bus bodies, motor car bodies and rebuilt Bristol *Fighters*. Commercial and sporting aviation was still in the future and besides attempts to sell *Fighters* as *Tourers* a succession of prototypes were built.

In mid-1920 Bristol had acquired the Cosmos aero-engine works at Fishponds, Bristol which was moved

Above:
SIR GEORGE WHITE Bt. 1854-1916
Bristol-born millionaire entrepreneur and founder of the British & Colonial Aeroplane Co. Ltd., established at Filton, Gloucestershire in February 1910.

Left:
BRITAIN FIRST
The Type 142, named 'Britain First' by its owner Lord Rothermere, was Bristol's first 'modern' aircraft and is seen here in 1935 bearing RAF markings and serialled K7557. *Aeroplane.*

to Filton to become the Bristol Engine Department. Sales of the *Jupiter* 9-cylinder radial engine and a worldwide network of licencees brought in valuable funds and sustained Filton in the difficult decade up to 1930.

Bristol Aeroplane Company was unique in being able to supply complete aircraft and amassed considerable experience in all aspects of aircraft manufacture.

Aluminium alloys were still undeveloped and the principal forms of construction had used high tensile steel tubing and wood for the aircraft framing. In the mid-'twenties Harry Pollard came to Bristol and was responsible for the development of a patented method of fabricating airframes using high tensile steel strip. A rolling mill was set up at Filton to form the sections required for various parts of the airframe.

The first Bristol machine to feature this method of construction was the *Bloodhound (Fighter C)* which together with the *Beaver, Boarhound* and *Badminton* gave valuable experience.

The renowned Bristol *Bulldog* used a steel strip airframe covered in doped fabric and was chosen as the front line fighter for the RAF. Powered at first by the Bristol *Jupiter* it was proposed to repeat the sales success with a *Bulldog* having a new Bristol engine, the *Mercury*. Unfortunately the RAF preferred the *Mercury* engined Gloster *Gauntlet* instead but Bristol at least gained half the sale.

Bulldogs for the RAF and for export were the mainstay of the airframe works in the early 1930s but the Company's profit came mostly from the sale of aero engines.

Above:
BRISTOL BIPLANE
Captain Dickson's army manoeuvres on Salisbury Plain in September 1910. The Biplane ("Boxkite") is No.9 with 50hp Gnome rotary engine.

Below:
COANDA
Coanda monoplane at Larkhill on Salisbury Plain in 1912. The early 'British & Colonial' types were sold under the brand name "The Bristol" to avoid problems in registering the name 'Bristol' as a trade mark.

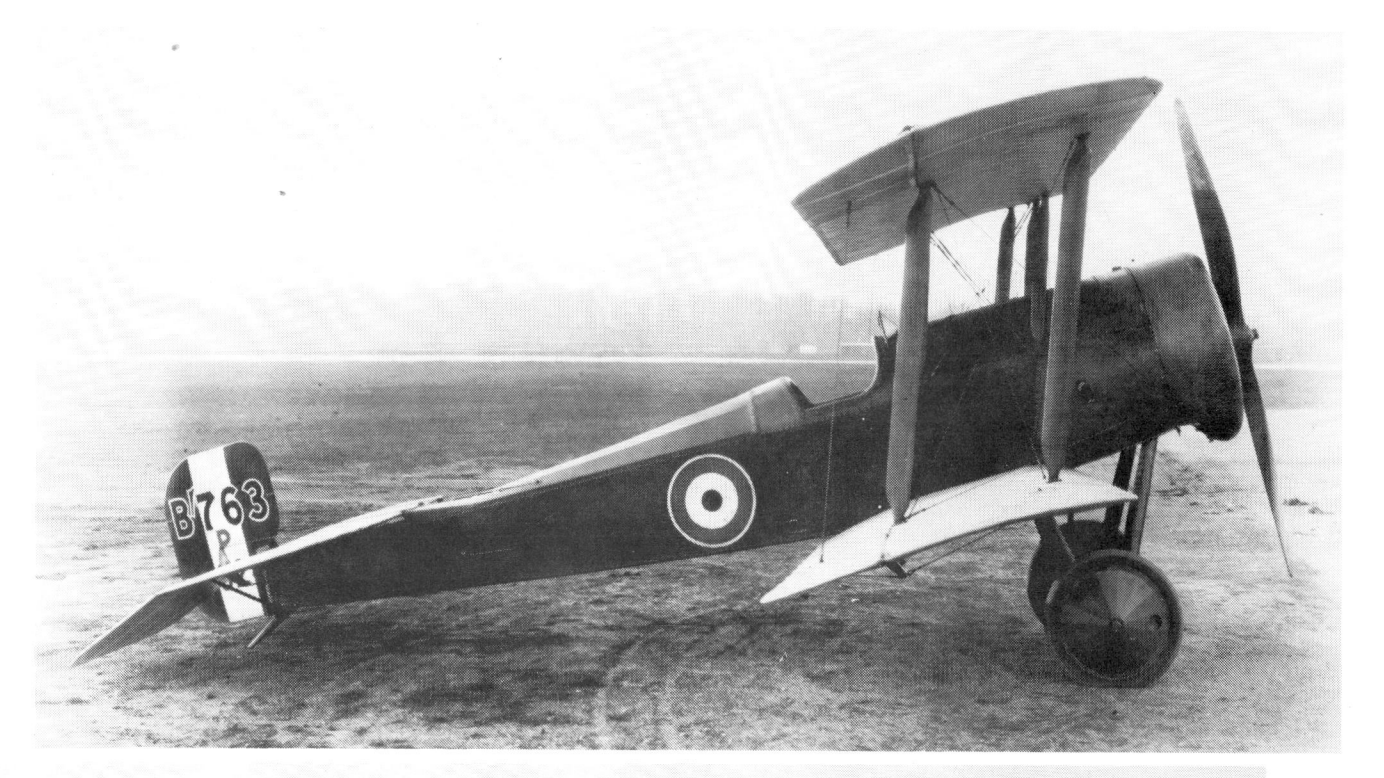

Top:
BRISTOL SCOUT
A Scout D rebuilt by No.1 (Southern) Aeroplane Repair Depot. It was sent to McCook Field, Dayton, Ohio for evaluation where it received the identify P32 in addition to the RFC serial as shown here.

Above:
BRISFIT
The most famous British & Colonial machine was the Bristol Fighter. J8257 was a Mk.IV conversion with dual control and Handley Page auto-slats.

Below:
JUPITER RACER
The Bristol Jupiter Racer G-EBDR had just sufficient airframe to enable the Jupiter engine to fly. In flight it was barely controllable and only one was built but is worthy of inclusion because it was the first aircraft in the world to have a retractable undercarriage, albeit hand operated.

Above:
BRISTOL BULLDOG
The prototype Bristol Bulldog II serial J9480 with Bristol Jupiter VII engine. Serving as the RAF's front line fighter in the early 1930s, the Bulldog was Bristol's most successful biplane of the interwar years. Apart from the raw materials, the only parts that were not built at Filton were the wheels, tyres and wireless set.

Below:
BRISTOL BOMBAY
L5808, the first of fifty production Bombay transports built by Shorts in Belfast because Filton was fully occupied with work on the Blenheim bomber.

Filton discarded biplanes after 1933 and attempts were made to develop single engined fighter monoplanes and twin engined monoplane transports.

In the single engined type Bristol insisted on using its own radial engines but it was a hard battle against the water cooled Rolls Royce engines with their smaller frontal area. Eventually it became a lost cause but this was partly due to Bristol's success with twin engined aircraft.

Bristol had flown the *Bagshot* in 1925 but knowledge of large monoplane wings was limited and this twin engined monoplane fighter remained an experiment.

After several schemes had been worked out by Barnwell and his team, a low wing cantilever monoplane having two Bristol *Mercury* engines was built in 1934-35 to the order of Lord Rothemere, millionaire newspaper proprietor and aviation enthusiast. Known as the type 142 it was christened *Britain First* by its owner and presented to the Nation as an example of a fine modern aircraft. A similar machine, type 143, with two Bristol *Aquila* sleeve valve engines, was built concurrently with the type 142. Both machines were of stressed skin aluminium construction and their importance to Filton cannot be overstated.

Official interest in the *Britain First* resulted in a development of it as a bomber for the RAF and the prototype type 142M Blenheim (K7033) first flew on 25 June 1936. It was immediately ordered into production and steps were taken to prepare the airframe and *Mercury* engine for 'shadow' production at factories dispersed throughout England.

A year previously on 15 June 1935 the firm had been reformed as a public limited company with a share capital of £1,200,000. This coincided with an upsurge in demand for military aircraft to meet the threat of Hitler's build up of the Luftwaffe.

Once the *Blenheim* had gone into production other projects took a back seat in official circles and the type 146 single seat fighter and type 148 army co-operation monoplane remained solely as prototypes.

A larger twin engined bomber known as the *Bolingbroke* was proposed in 1937 but emerged two years later as the 'long nosed' *Blenheim IV*. It superseded the original 'short nose' *Blenheim I* in shadow production. Examples of both types were converted to fighters with guns under the nose and black-painted *Blenheim I*'s saw action as night fighters until replaced by the *Beaufighter*.

A derivative of the *Blenheim*, the type 152 *Beaufort* was built as a torpedo bomber for RAF Coastal Command. The first *Beaufort (L4441)* was flown on

Above:
BRISTOL BLENHEIM
K7138 from the first batch of 150 Blenheim I's built in 1936/37. Powered by two 840hp Bristol Mercury VIII engines, over 5,000 were built by Bristol and contractors and 'shadow' production included 1,000 Blenheims built by A.V. Roe at Chadderton and Woodford.

15 October 1938 but due to overheating problems with its new two row Bristol *Taurus* engines did not enter RAF service until January 1940.

A further descendent of the *Blenheim* was known as the type 156 *Beaufighter* which was powered by two Bristol *Hercules* 14-cylinder radial engines. When the new radar devices being developed became too heavy and cumbersome to fit into the wings of a single engined fighter the *Beaufighter* was chosen as the RAF night fighter. After the 'Battle of Britain' the need for home based night fighters diminished and the *Beaufighter* was then built to suit the needs of RAF Coastal Command. *Beaufighter TFX* torpedo fighters, the main form of the type 156, were built at Filton and Weston-Super-Mare while other marks were built by Fairey Aviation, Rootes and in Australia.

Schemes for a 'Beaubomber' known as the *Beaumont* did not come to fruition but evolved as the type 163 *Buckingham* with two 18-cylinder Bristol *Centaurus* engines. *Buckinghams* came too late to see wartime service as bombers and most were converted to light courier transports or *Buckmaster* trainers.

Bristol's final wartime design was the type 164 *Brigand* torpedo bomber using the wings and tail of the *Buckingham* married to a new fuselage. *Brigands* were also too late for the war but saw action in the Far East against terrorists in the late 1940s.

Below:
BRISTOL BEAUFORT
EK982, an early production Beaufort I posed outside No.1 Flight Shed at Filton in 1942. The Bristol Taurus engined Beaufort was a derivative of the Blenheim for torpedo bombing and served with RAF Coastal Command until replaced by the Beaufighter TF.X.

Above:
BRISTOL BEAUFIGHTER
Beaufighter IF serial R2198 with two 1,400hp Bristol Hercules engines. The Beaufighter became the definitive RAF night fighter when equipped with airborne interception radar. Besides the production lines at Filton and Weston-Super-Mare, Beaufighters were also built by Fairey Aviation at Stockport, Cheshire and by Rootes at Blythe Bridge, Staffs., as well as in Australia

Below:
BRISTOL BRIGAND
Brigand B.I (RH806) built at Filton in 1948 and powered by two Bristol Centaurus 57 engines. Tropicalised versions of the Brigand light ground attack bomber saw active service in the Far East against Malayan terrorists.

A corporate re-organisation occurred at Filton in 1944 when the various departments gained divisional status. These were the Aircraft Division covering the old works while the Engine Division occupied the East and West Works straddling the A38 trunk road. An Armaments Division took care of Bristol power operated turrets and this formed the basis of the guided weapons activity from 1949. A Housing Division at Weston Super Mare built aluminium houses after aircraft work tailed off at war's end. To diversify, a prestigious range of two litre sports cars was started in 1946 with the Bristol 400.

During the war Bristol had proposed a 100 ton bomber using eight *Centaurus* engines but it was not built. When the Brabazon Committee published its list of types of post war civil transports Bristol received a Ministry of Supply contract for two trans-Atlantic airliner prototypes, based on the 100 ton bomber project. *Brabazon 1 Mk.1* was flown for the first time on 4 September 1949 powered by eight *Centaurus* engines but the second prototype did not fly because its Bristol *Proteus* turboprops were diverted to the SARO *Princess (G-ALUN)*. When the Tory Government returned to power in 1953 the *Brabazon* was cancelled.

A smaller commercial aircraft had sustained the airframe works at Filton in post war years. This was the type 170 *Freighter* and its passenger equivalent the *Wayfarer*. First flown on 2 December 1945 the type 170 was built at Filton until 1957. Of the 214 aircraft that were built at least one *(G-BISU)* is still flying in revenue service in this country plus two in Canada.

Work on the *Britannia* airliner to meet another of the Brabazon Committee's types commenced in 1948 when three prototypes were ordered by the Ministry of Supply. It was intended that these should have Bristol *Centaurus* piston engines but the first Britannia, *G-ALBO*, flew with Bristol *Proteus* turboprops on 16 August 1952.

Fifteen *Britannia 102's* were sold to BOAC and were delivered from December 1955 onwards, entering service on the Johannesburg run in February 1957. Subsequent orders were for the 'stretched' version in medium or long range forms which included a cargo version for the RAF.

Above:
BRISTOL BRITANNIA
G-AOVB, the first Britannia 312 shows the graceful lines of Bristol's "Whispering Giant". Icing problems with its four Bristol Proteus turbo prop engines delayed its entry into service with BOAC and restricted its sales to 75 aircraft.

Below:
BRISTOL FREIGHTER
The third Freighter Mk.1 (G-AGVC) as converted to Mk.31 with larger tail fin. Freighters with nose doors and the passenger version known as the Wayfarer sustained the Filton aircraft works in the immediate post war years.

Further changes in the corporate structure during 1956 saw the divisions elevated to subsidiary companies. Bristol Aircraft Limited, Bristol Aero Engines Limited and Bristol Cars Limited became wholly-owned subsidiaries of Bristol Aeroplane Company Limited. All helicopter work was transferred to Weston and the Housing Division became the Helicopter Division until merged with Westland in 1960.

The aircraft most important to the longevity of Filton were the type 142 *Britain First* and the *Brabazon*. *Britain First* was Bristol's first 'modern' aircraft and the large production orders for derivatives of it (*Blenheim, Beaufort* and *Beaufighter*) gave Filton considerable experience in monocoque light alloy construction. The 'shadow' factory scheme taught Bristol the skill of unit construction and the value of interchangeability.

Only one Brabazon was flown and the project was a commercial failure but Filton inherited two extremely valuable assets from *Brabazon I*. Without the Assembly Hall and long runway built for the 'Brab' Filton might not have survived the traumas of the 1960s.

With the emergence of jet airliners the *Britannia* was not the success that was anticipated and the icing problems with the *Proteus* engine placed a severe financial burden on the Company through lost sales. As a result Bristol Aeroplane Company was in a weak position to resist Government pressure to merge its engine interests with Armstrong Siddeley Motors in 1959 (so forming Bristol Siddeley Engines) and its airframe works with English Electric and Vickers Armstrongs (British Aircraft Corporation) in 1960.

Bristol Aeroplane Company continued to exist after 1960 as a parent of British Aircraft Corporation until it was taken over by Rolls Royce in 1966 in order to gain control of Bristol Siddeley. The name finally became extinct when Rolls Royce went bankrupt on 4 February 1971.

Above:
BRISTOL 188
XF926, the second type 188 stainless steel research aircraft pictured in July 1963. Powered by two de Havilland Gyron Junior turbojets it was the last fixed wing Bristol aircraft to be built.

Left:
BRISTOL BLOODHOUND
Developed under the code name 'Red Duster', the Bristol/Ferranti Bloodhound surface to air missile entered RAF service in 1957. Bloodhounds are accelerated to supersonic speed from rest by four jettisonable solid rocket boosters and then powered to target by two Bristol Thor ramjet engines.

Above:
BRISTOL 173
The third prototype 173 helicopter, XE286, built at Filton with two Alvis Leonides Major engines seen hovering at Oldmixon near Weston Super Mare in November 1956. A later version, type 192 Belvedere, entered service with the RAF and was the last Bristol type to be built, albeit under Westland management.

Below:
FILTON
The Bristol Aeroplane Company works at Filton near Bristol (now British Aerospace, Bristol) photographed in October 1983. The A38 trunk road is in the foreground.

DE HAVILLAND

DH

When writing about de Havilland it is fitting to follow the progress of the founder not just the company that bore his name. Geoffrey de Havilland was born in Hampshire on 27 July 1882. The son of the Rev. Charles de Havilland he chose not to join the family 'business' but to seek a career of his own in mechanical engineering.

After several years as a 'bus designer in London he sought to satisfy his enthusiasm for the new science of aviation by designing a rudimentary biplane. The construction of a small engine of his own design was contracted to the Iris Motor Company at Willesden, London but the airframe he built himself, with help from his brother-in-law Fred Hearle, in a rented workshop in Fulham. The machine was taken by road to the North Hampshire Downs in 1909 but was wrecked after only a very short 'flight'.

Undaunted, de Havilland set to building a second machine, a pusher biplane similar to a Farman boxkite in which he installed the engine from the first biplane. It was successfully flown in the hands of its creator on 10 September 1910 and was subsequently sold to the War Office which designated it F.E.1. (Farman Experimental).

1912 saw the formation of the Royal Flying Corps and the renaming of the Balloon Factory as the Royal Aircraft Factory to encompass heavier than air machines as well as the lighter ones. During his two year stay at Farnborough after joining the RFC de Havilland was involved with the design of the S.E.1. and helped to create the B.E.1.; progenitor of a series of B.E. tractor biplanes.

George Holt Thomas, wealthy newspaper proprietor and aviation enthusiast, formed The Aircraft Manufacturing Company Limited in 1912 at The Hyde, Hendon, North London. Geoffrey de Havilland joined Air-Co, as it was more familiarly known, in June 1914 as chief designer.

He proceeded to design a series of gunbus type pusher biplanes which were named after the designer in the manner of the times. The Air-Co D.H.1 powered by a 70hp Renault engine was flown in 1914 and the production version, D.H.1A., with a Beardmore engine was built by Savages in Norfolk. A smaller single-seat version known as D.H.2. with Gnome or Le Rhone rotary engine was built by Air-Co for the RFC.

Using two engines for the first time in a de Havilland machine, the D.H.3. was a biplane bomber which was unusual for the time in having fold-back wings for stowage. Only two were built.

In contrast, over six thousand of the next type, D.H.4. were built by Air-Co and subcontractors. This was a single engined day bomber first flown in 1916 and which due to the urgency of war appeared in many versions.

The pattern was repeated for the next and last five types to be produced under the Air-Co banner and apart from the D.H.5. of which only 550 were built all exceeded a production run of one thousand.

With the ending of the war production quantities were much reduced and the D.H.11, D.H.14, D.H.15 and D.H.18 only totalled eleven aircraft altogether. These were sold under the trade name of Airco.

Threatened by Excess Profits Duty and with few peace time prospects, Holt Thomas sold out to Birmingham Small Arms Company Limited (BSA) in March 1920.

Geoffrey de Havilland had proven himself as a major force in British aviation and this gave him confidence to branch out on his own. With some financial help from George Holt Thomas The de Havilland Aircraft Company Limited was incorporated on 25 September 1920. A sound design, engineering and management team was recruited from Airco and these men including F.T. Hearle became directors of the new venture. This select team stayed together throughout most of the life of de Havilland Aircraft.

Premises were rented in Stag Lane, Edgeware, then a quiet rural district. The airfield had formerly been used as a flying school by the London & Provincial Aviation Company.

The first aircraft type to be built by the new company, the D.H.27. long range day bomber biplane, was actually conceived at Airco. In contrast to the heady days of the Great War when every de Havilland design went into large scale production, the 'twenties saw more proposed aircraft than real ones.

Above:
GEOFFREY DE HAVILLAND 1882-1965
The founder of the de Havilland empire pictured circa 1913 while serving in the Royal Flying Corps Special Reserve as a second lieutenant.

Opposite:
DRAGON RAPIDE
F-BEDY was photographed at de Havilland's Witney, Oxfordshire base in 1948 prior to delivery to Air France.

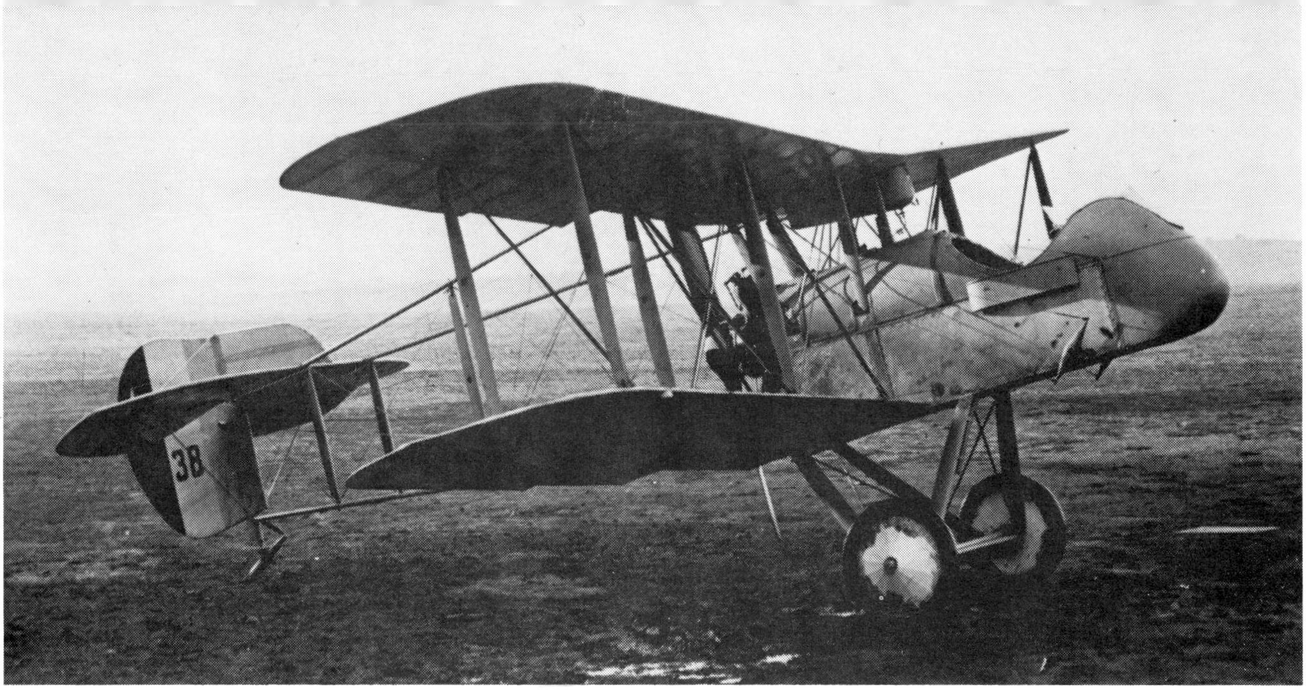
38

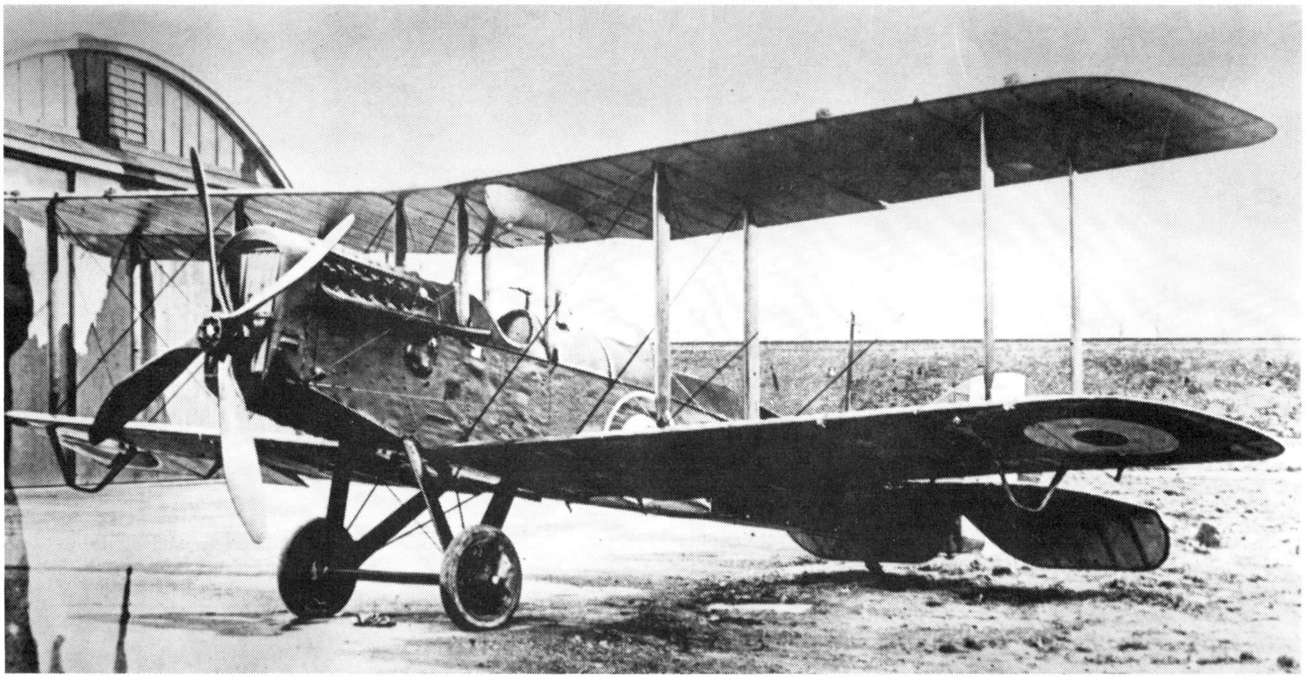

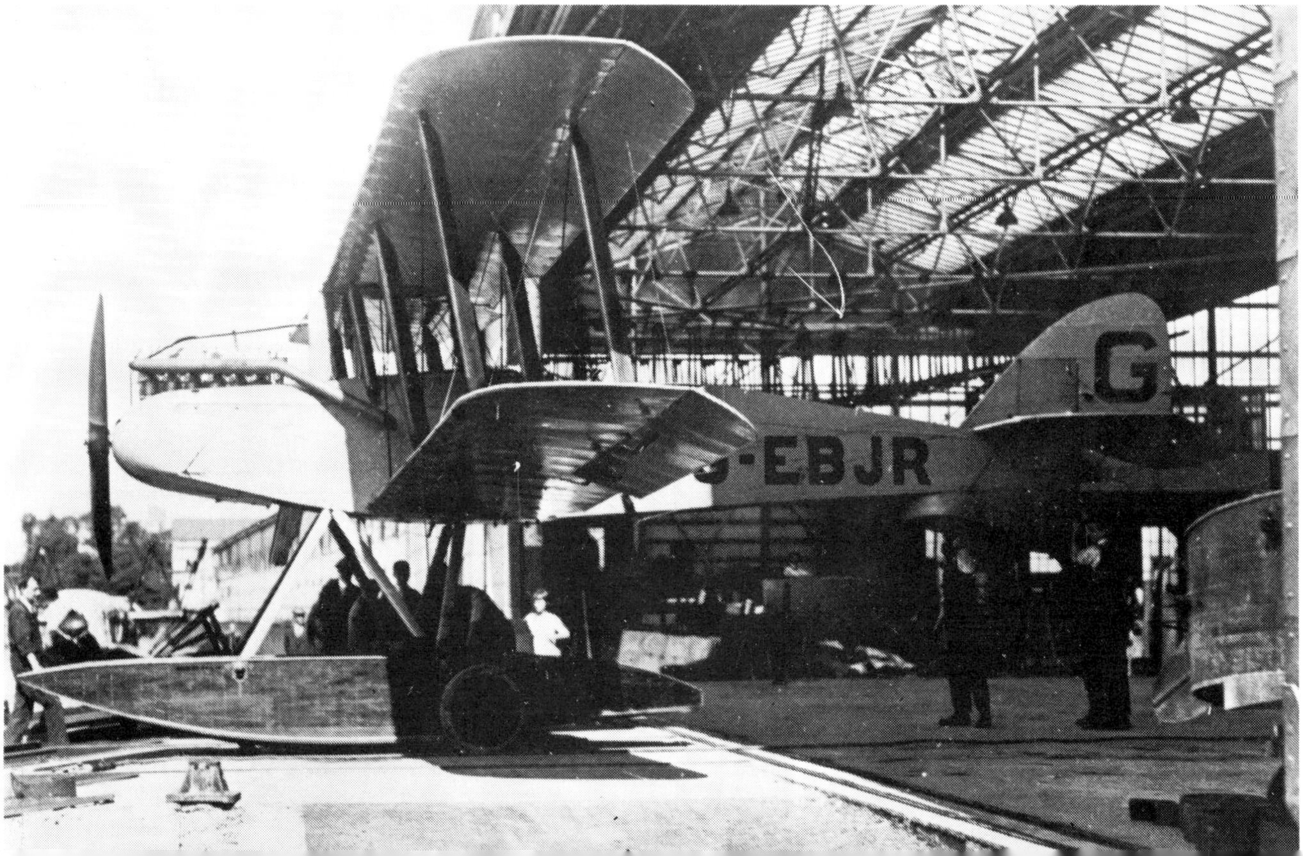
G
-EBJR

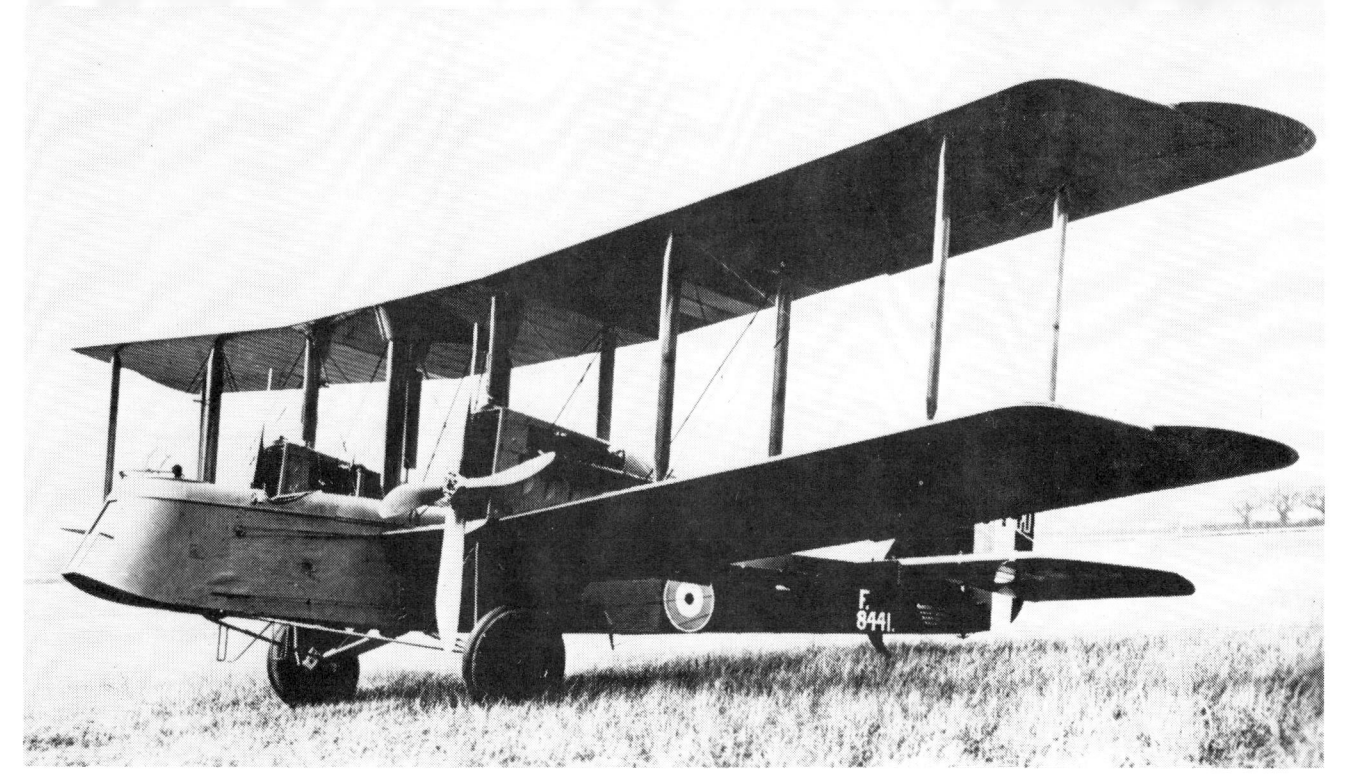

Left Upper:
AIR-CO D.H.2
The D.H.2 was the first of Geoffrey de Havilland's designs to be built in large quantity. 452 were built by the Aircraft Manufacturing Company Ltd (Air-Co) at Hendon, north of London. No.5938 shown was itself built in 1915.

Left Centre:
AIR-CO D.H.4
A7995, shown here, was built by Air-Co in 1916 and had a BHP engine. The D.H.4 day bomber was the first de Havilland design to be built in very large volume; over 6,000 being built with various engines by Air-Co and licencees in the UK and USA.

Left Lower:
AIR-CO D.H.9
G-EBJR shown here in its 1924 condition with Armstrong Siddeley Puma engine was built as H9289. Over 6,000 D.H.9 and D.H.9A's were built by Air-Co and fifteen licencees during the Great War.

Above:
AIR-CO D.H.10A
The D.H.10 Amiens was de Havilland's first successful twin engined aeroplane. F8441 illustrated was built under licence by Mann-Egerton at Norwich.

A radical change of ideas took place with the building of the D.H.29 *Doncaster* which was a high wing cantilever monoplane powered by the trusty Napier *Lion*. Designed as a ten seat civil transport it first flew in 1921 and was used experimentally but did not go into passenger service.

The years 1920 to 1925 were unsettled and difficult for all manufacturers; no less de Havilland. However, the Company did not diversify into non-aircraft products and persevered with aircraft. Being a small company with low overheads undoubtedly helped in this respect.

Military orders were scarce and only five Service aircraft were built in the period. They were the D.H.42 *Dormouse* two seat reconnaissance biplane with an A.S. *Jaguar* engine of which two were built and a relative, the Bristol *Jupiter* engined D.H.42A *Dingo* army cooperation biplane. The D.H.56 *Hyena* was also for army cooperation duties.

De Havilland was more fortunate in the civil aircraft market starting with the D.H.34, a *Lion* engined nine-passenger cabin biplane of which eleven were built and operated by Instone Air Lines and Daimler Hire.

Intended as a D.H.9 replacement, the later D.H.50 made good sales for the period. The prototype D.H.50 first flew in August 1923 and was succeeded by the slightly larger D.H.50A which was built under licence in Australia, Belgium and Czechoslovakia. QANTAS operated D.H.50J's powered variously by *A.S. Jaguar* and Bristol *Jupiter* engines until 1937. The D.H.54 *Highclere*, named after Geoffrey de Havilland's childhood home village in Hampshire, was a twelve passenger cabin monoplane designed to an Air Ministry specification and using a Rolls Royce *Condor* engine. Only the prototype, *G-EBKI*, was built.

In the early 1920s the *Daily Mail* newspaper organised aeroplane competitions to boost its circulation. For its part, de Havilland offered the D.H.53 *Humming Bird* using a Douglas motor cycle engine and later a Blackburne *Thrush* and Bristol *Cherub*. The prototype was flown in the 1923 trials at Lympne and the interest in it generated total sales of fifteen aircraft.

The D.H.51 was an important aircraft in the history of de Havilland for although only three were made it marked the turning point in the Company's fortunes. Powered by a war surplus RAF.1A. engine it was the first attempt to produce a rugged private owner aeroplane at a sensible price. *G-EBIR*, the third D.H.51 which was built in 1925 is now preserved by the Shuttleworth Trust.

A conscious decision was made in the mid-1920s to concentrate on the civil market particularly the private owner and sporting type. With light aircraft de Havilland was to excel and from the D.H.60 *Moth* which was first flown on 25 February 1925 a succession of types was produced in healthy quantities; in hundreds not thousands but in those difficult years a sale of fifty aircraft was worthwhile. Production of the eight versions of the D.H.60 *Moth* did in fact reach 2112 aircraft all told.

All the D.H.60 series were biplanes but were followed by a short period of experiment with high speed monoplanes including the D.H.71 and D.H.75.

A boom in private house building in the latter part of the 1920s was also noticeable at Stag Lane Works which became hemmed in by a vast expanse of new houses. To gain breathing space a green fields site was bought at Hatfield, Hertfordshire in 1930. Airframe work had been transferred to Hatfield by the end of 1932 and the last aircraft to fly from Stag Lane did so two years later. Thereafter, Stag Lane became the de Havilland engine works building the *Gipsy* family of air cooled engines.

Before vacating the old works a start had been made on two aircraft types that were to be among the three big successes of the inter-war years; namely the *Tiger Moth* and the *Dragon* series. The D.H.82 *Tiger Moth* followed the same basic formula as the D.H.60T *Moth Trainer* but was further refined by using the inverted *Gipsy III* engine. Full production started in 1931 and was soon re-established at Hatfield. Adoption by the RAF of the *Tiger Moth* as a primary trainer to equip the rapidly expanding training scheme ensured large orders. Over 8,000 *Tiger Moths* were built altogether by de Havilland at Hatfield, Morris Motors, de Havilland factories in Canada, Australia and New Zealand, and in Sweden and Portugal.

The D.H.84 *Dragon* was built to the order of Hillman Airways for cross-Channel services and was to herald de Havilland's second most successful series of machines built between the wars. Having two de Havilland *Gipsy Major* engines the first *Dragon* flew at Stag Lane on 24 November 1932. *Dragons* became popular with small commercial operators and over two hundred were built including eighty-four by de Havilland in Australia.

For longer haul use in Australia the D.H.86 biplane was conceived with four de Havilland *Gipsy Six*

Below:
D.H.60 MOTH
G-EBLV with ADC Cirrus engine was just one of the highly successful two-seat touring biplanes built by the de Havilland Aircraft Company at Stag Lane in 1925. It crashed in 1929 but was restored in 1961.

Bottom:
D.H. 82A TIGER MOTH II
R5130 was built for the RAF just prior to the outbreak of the Second World War.

Right:
D.H.89A DOMINIE
The Dominie was a military variant of the Dragon Rapide and was powered by two de Havilland Gipsy-3 engines. X7398 leads X7386 in this pre-war portrait.

engines mounted on the lower set of highly tapered wings. The prototype flew at Stag Lane in January 1934 and sixty-two had been sold by the time that construction ceased in 1937.

To replace the *Dragon* a two engined version of the D.H.86 was introduced and, named *Dragon Rapide,* it first flew at Hatfield on 17 April 1934. More usually known as the *Rapide,* it soon surpassed the sales record of the D.H.84 but in this respect it proved to be something of an embarrassment. With the increased workload for military aircraft production, it had to be subcontracted to Brush Coachworks in Leicester in 1943 where almost half of the 727 *Rapides* were built.

To cash in on the popularity of the *Dragon* series a well appointed private owner version, the D.H.90 *Dragonfly* was a scaled down version of the *Rapide* with structural differences. It found favour as a light communications aircraft and sixty-seven were built.

While other companies like Avro, Bristol and Vickers had given a great deal of time and effort to all metal construction, de Havilland specialised in wooden or composite wood and metal airframes. The D.H.88 *Comet* racer of 1934 had given a valuable lesson in wooden construction and further experience with subsequent types built up a high level of knowledge in monocoque wooden construction at Hatfield.

In 1936 Hatfield received an order for two trans-Atlantic mailplanes for experimental purposes and the prototype first flew at Hatfield on 20 May 1937. Using a sandwich of cedar plywood interleaved with a larger thickness of balsa a strong monocoque structure could be built up in sections on jigs. Named D.H.91 *Albatross*, they were flown experimentally on mail routes by Imperial Airways crews to Egypt and Pakistan.

When war was declared they were impressed into military service as long range courier transports.

Imperial Airways ordered five *Albatrosses* as 122 passenger transports and the first was delivered in 1938. Due to crashes, enemy action and shortage of spares they did not survive the war.

Based on the experience with the *Albatross*, de Havilland proposed in 1938 a wooden bomber using two Rolls-Royce *Merlin* engines. Official response to the idea was cool but when the shortage of strategic metals was realised the project gained momentum.

Geoffrey de Havilland jnr., took the prototype *Mosquito* into the air for the first time on 25 November 1940. Trials in 1940 and '41 showed that the *Mosquito* was better than expectations and it went into service in 1942.

Originally conceived as a light bomber the *Mosquito*, on account of its high speed, found favour as a nightfighter and for photo-reconnaissance duties. Later versions were equipped as dual control trainers, fighter bombers armed with rocket projectiles and inevitably as target tugs. A *Sea Mosquito* was devised for torpedo laying and reconnaissance work aboard aircraft carriers.

To meet the demand for increased wartime production the capacity of outside firms was enlisted as had been the case in the Great War. Leavesden was established as a shadow factory and became known as de Havilland factory number two. Airspeed at Christchurch and Portsmouth, under de Havilland control from 1940, was well suited to *Mosquito* production. Work was also subcontracted to Percival Aircraft at Luton, Standard Motors at Coventry and at Hawarden near Chester. Overseas, de Havilland Aircraft of Canada built *Mosquito* fighters, bombers, fighter-bombers and trainers while the Australian subsidiary turned out fighter-bomber, trainer and photo-reconnaissance versions.

Altogether, 7,659 "Mozzies" were built and gave sterling service in all theatres of war but are probably best known for their pathfinding role. On difficult to find enemy targets the pathfinding "Mozzies" would mark the aiming point with flares which could easily be found by the main force of *Lancasters*.

Top:
D.H.95 FLAMINGO
G-AFYH which saw active service with the Royal Navy as BT312 is seen here flying in post war condition. The Flamingo was de Havilland's first all-metal stressed skin monocoque airframe and was powered by two Bristol Perseus XIIC sleeve valve radial engines.

Above:
D.H.98 MOSQUITO
Hatfield built Mosquito B.XVI, serial ML963. The Rolls Royce Merlin engined "Wooden Wonder" otherwise known as the "Mozzie" became famous for its pathfinding and photo-reconnaissance roles in World War Two. Another Mosquito RR299 has been restored by voluntary helpers and resides at British Aerospace, Broughton near Chester.
MoD.

To complete its range of products de Havilland became involved in propeller manufacture and obtained a licence to produce the American Hamilton propeller. From 1935 they were made at the engine works in Stag Lane but on the formation of de Havilland Propellers Limited in 1946 work was centred in a new building at Hatfield. Facilities at Stevenage and Lostock, near Bolton, were also acquired for propeller work.

Engine production increased steadily and The de Havilland Engine Company Limited was formed in 1944. Having its own aero-engine expertise gave de Havilland the key to post-war development and production of jet propelled aircraft both large and small. Major F.B. Halford designed a turbojet having a centrifugal compressor and this was built by de Havilland as the *Goblin*. To exploit the potential of this engine the D.H.100 *Vampire* having a stubby fuselage and twin-boom tail was devised.

Geoffrey de Havilland jnr., flew it on 29 September 1943 and it was almost immediately put into production at English Electric. Nearly three thousand *Vampires* were produced in the early post war years and besides those built by Hatfield, Samlesbury and Manchester, *Vampires* were built under licence in Switzerland, Italy, France and as far away as India.

From the *Vampire* was developed the D.H.112 *Venom* single seat fighter bomber with more powerful *Ghost* engine which first flew at Hatfield on 2 September 1949. *Venoms* were built in fighter, fighter-bomber, all weather fighter and night fighter versions.

The D.H.113 *Vampire N.F.* (night fighter) had side by side seating for a crew of two and from this was developed the D.H.115 *Vampire Trainer*.

Success with military jet aircraft did not lull the Company into neglecting the potential civil market. From the Brabazon Committee's list of recommendations Hatfield selected the type 4 for which it proposed the D.H.106 *Comet* and the type 5B which gave rise to the *Dove*.

The all metal *Dove* was started before the war had ended and the prototype, *G-AGPJ*, was flown on 25 September 1945 to coincide with the 25th anniversary of the formation of the Company. It attracted sales to airlines servicing short haul routes and in 1948 the *Dove 2* executive aircraft was introduced. *Dove* production ceased at Hatfield in 1951 but continued at Hawarden.

While other British manufacturers were involved with turboprop airliners de Havilland took the bold step of building the turbojet powered *Comet*. Bearing the 'class B' mark *G-5-1* the prototype *Comet* flew from Hatfield on 27 July 1949. Ten *Comet 1's* with *Ghost* engines were delivered to BOAC in 1952 and the first of these, *G-ALYP,* inaugurated the world's first pure jet passenger service on 2 May 1952.

All went well until 1953-54 when the unexplained loss of three *Comets* caused the fleet to be grounded. Delays were inevitable while engineering changes were made to resolve the fatigue problems around the window apertures.

To meet BOAC's demand for a trans-Atlantic airliner a 'stretched' *Comet 2* went into production but never went into passenger service and were diverted to the RAF. Only one *Comet 3* was built and long range *Comet 4's* went into BOAC passenger service between London and New York on 4 October 1958. Besides being the world's first pure jet airliner the *Comet* pioneered the use of Redux bonding of aluminium in large structures and the process has been used extensively by successor companies including British Aerospace.

Above:
D.H.104 DOVE
 G-AOSE with two de Havilland Gipsy Queen engines in the service of Smiths Industries Limited based at Staverton, Glos. Built as a Dove 6 circa 1956 it survived until 1974.

Below:
D.H.106 COMET
 G-ARDA the first of nineteen Comet 4 airliners ordered by BOAC which went into service on North Atlantic routes in 1958. It was sold to Malaya in 1965 (9M-AOA) but returned to the UK in 1969.

Above:
D.H.110 SEA VIXEN
XJ578, a Sea Vixen F.A.W.I., was built at the Airspeed factory at Christchurch, Hants., after it had been taken over by de Havilland.

Below:
HATFIELD
BAe Hatfield (formerly the de Havilland Aircraft works since 1930) looking north west with the A1 (M) bypass under construction. In this 1986 view the former de Havilland Propellers and DH Engine works (later H.S. Dynamics) lies on the far side of the runway.

The penultimate de Havilland type, the D.H.121 *Trident*, was born amid a time of turmoil in the British aircraft industry.

In 1956, British European Airways Corporation issued its requirement for a short range transport having turbojet engines giving a speed of about 600mph. The aircraft had to be able to take-off and land from smaller paved airfields and be made such that it could be serviced and 'turned round' quickly to give maximum utilisation.

To meet this specification de Havilland proposed the D.H.121 against fierce competition including the Bristol 200. Duncan Sandys, champion of the Tory Government used the BEA order as a weapon to enforce an amalgamation within the industry. Design and manufacture of the D.H.121 was originally intended to be accomplished by a consortium of Hunting, Fairey and de Havilland for which the group name Airco had been revived. Under Government pressure The de Havilland Aircraft Company Limited, de Havilland Propellers and de Havilland Holdings Limited became part of Hawker Siddeley Group in 1960. The BEA aircraft therefore became solely a de Havilland effort. De Havilland Engines was absorbed by Bristol Siddeley Engines in 1961.

Named *Trident* in 1960 the D.H.121 powered by three Rolls Royce *Spey* turbojets was flown on 9 January 1962 and deliveries to BEA commenced in the middle of 1963. Twenty-three *Trident I*'s entered service with BEA and the type was further developed as the Hawker Siddeley HS 121.

At the smaller end of the scale Hatfield conceived the D.H.125 business jet using two rear mounted Bristol Siddeley *Viper* turbojets. The prototype G-ARYA flew on 13 August 1962 and on the formation of Hawker Siddeley Aviation in 1963 became the HS 125 and latterly BAe 125.

Above:
D.H.121 TRIDENT
G-ARPC was the third of twenty-four Trident IC's built for British European Airways and first flew on 25 August 1962. Besides its three Rolls Royce Spey propulsion engines it also had a Bristol Siddeley Artouste auxiliary engine/generator in its tail to supply power for ground services.

Below:
D.H.125
Prototype D.H.125 executive jet, G-ARYA, powered by two Bristol-Siddeley Viper-20 turbojets. The D.H.125 was the last de Havilland design but actually made its maiden flight after Hatfield had become a part of Hawker Siddeley Group in 1960.

Blackburn
AIRCRAFT

The history of Blackburn Aircraft (now part of British Aerospace at Brough) was the result of the enthusiasm of one man. Robert Blackburn was born on 26 March 1885 and was brought up in Yorkshire where he gained an honours degree at Leeds University. After travelling through Europe he joined the Leeds firm of Thomas Green & Sons Limited where his father was General Manager.

Bob Blackburn's first attempt at building a flying machine in a room at Benson Street, Leeds in 1909 met with failure. His early monoplane built on a tricycle type underframe and powered (underpowered?) by a 35hp Green engine was unable to lift off the sands at Marske-by-the-Sea, Yorkshire.

His second attempt met with complete success. This was also a monoplane but resembled the French Antoinette with a triangular section fuselage. Using an Isaacson engine of some 40hp it achieved first flight at Filey in 1911.

Robert Blackburn established himself in a bungalow on the cliffs at Filey and nearby was a slipway leading from a shed down to the sands. Here, in isolation, R.B. persevered with further development and produced the *Mercury* monoplane having a 50hp Isaacson engine. This machine was exhibited at Olympia in 1911 and others were erected using Gnome, Renault and Anzani engines.

To establish himself as an aeroplane constructor rather than just an enthusiastic amateur, R.B. acquired premises in Balm Road, Leeds in 1911. He had also formed Blackburn Aeroplane Company by this time.

In the following year a military aeroplane competition was held and Blackburn built a military version of the *Mercury* which has since become known as the Blackburn type E. It was not finished in time but gave much valuable experience and similar monoplanes were built at Balm Road for private buyers.

Three monoplanes were erected in 1913 with Gnome or Anzani engines for private buyers and one of these took part in the 'War of the Roses Race' held in that year.

A type L seaplane with 130hp Canton-Unne engine was built in 1914 for the 'Circuit of Britain' race and an order was received for a dozen B.E.2c biplanes designed by the Royal Aircraft Factory. Needing larger premises for this work the disused Olympia roller skating rink in Roundhay Road, Leeds was acquired in 1914.

On the eve of the Great War in 1914 the Yorkshire enterprise took a quantum leap forward when the Blackburn Aeroplane & Motor Company Limited was registered with a capital of £20,000.

On the outbreak of war the type L seaplane was impressed by the Navy and Blackburn was given orders for the licenced production of the Sopwith *Cuckoo*, Sopwith *Baby* and A.D.*Scout* in addition to further B.E.2c's. As well as 'foreign' aircraft, Blackburn also turned out several indigenous designs. These included the *Twin Blackburn (T.B.)* a naval biplane with two fuselages. Powered by two Gnome or Clerget engines and mounted on floats, nine examples were built for dropping exploding darts on marauding *Zeppelin* airships sent over by the Kaiser.

A small triplane of the gunbus type was built with a 130hp Clerget engine driving a pusher propeller but like the contemporary *G.P.* seaplane did not go into production.

The most important Blackburn type of the First World War, twenty being built in 1917, was the *Kangaroo* powered by two Rolls Royce *Falcon* engines. They were constructed as bombers but were converted to civil use after the war.

During 1916 Blackburn established a seaplane experimental department and aerodrome on the north shore of the river Humber at a little place called Brough. The *G.P.* seaplane was in fact the first Blackburn machine to be built at Brough in the Experimental Hangar. After the site was requisitioned by the Government other buildings were erected.

Due, perhaps, to the wartime experience and personal contacts built up during that time Robert Blackburn decided that his company should specialise in Naval aircraft. A flying boat (N.l.b.) had been partially built in 1917 and three *Blackburd* torpedo bombers were erected at Brough in 1918. This design was followed in 1919 by the *Swift*, powered by a 450hp Napier *Lion* engine.

The British Admiralty showed no interest in the *Swift* which was sold in twos and threes to the navies of Japan, Spain and USA.

The ending of the war brought hard times to the British Aircraft industry and to sustain the works at Roundhay Olympia the building of 'bus and motor car bodies was undertaken in the early 'twenties.

Above:
ROBERT BLACKBURN 1885-1955
R.B. photographed on 1 October 1941.

Left:
BLACKBURN BEVERLEY
XH123 was one of forty-seven Beverley C.1. heavy transports built at Brough for RAF Transport Command between 1954 and 1958. They were an enlarged version of the GAL60 and used Bristol Centaurus 173 engines.

BLACKBURN MONOPLANE
BLACKBURN
BLACKBURN MONOPLANE

B
9985

Left Upper:
BLACKBURN MONOPLANE
Powered by a 70hp V8 Renault engine, the two-seat Blackburn type E military monoplane was built in 1912. Having a tubular steel framed fuselage it was the first British aeroplane to have an all metal fuselage. A similar single seat machine had a 60hp Green engine and both machines were erected at Balm Road, Leeds.

Left Centre:
BLACKBURN MONOPLANE
Blackburn single seat monoplane with 50hp Gnome engine built in Leeds for Mr. C. Foggin in 1912. It is illustrated here in recent times and is the oldest British aircraft still capable of flight.

Left Lower:
BLACKBURN KANGAROO
The sixteenth Kangaroo in civil use while still bearing its military serial B9985. It was used by the Claude Grahame-White Air Service as pictured here at Brooklands on 31 May 1919. Twenty Kangaroos with Rolls Royce Falcon engines were built as bombers in 1917 and were the first Blackburn aircraft to be built in quantity.

Below:
BLACKBURN DART
N9806 was built at Brough in 1925 as part of a batch of 32 Dart torpedo carriers for the Fleet Air Arm. Based on the Swift, the Dart was the first Blackburn type to exceed 100 in production; 117 being actually built at Brough.

Meanwhile at Brough, Blackburn had persevered with his ambition to be a supplier of aircraft for the Admiralty. His confidence was rewarded in 1921 by a large order for the *Dart*, a descendent of the *Swift*. The *Dart* firmly secured Blackburn as a supplier of torpedo bombers for the Fleet Air Arm and was only superseded by yet another Blackburn type, the *Ripon* built from 1926 onwards. In the intervening years sixty-two Blackburn *Blackburn* fleet spotters had also been built together with several experimental types.

Mention should be made here of the *Bluebird*, a small single-engined biplane built in 1924 in worthwhile quantity for private buyers.

1926 saw the first flight of the *Iris*, Blackburn's first attempt at a large multi-engined flying boat. Design actually commenced to meet an Air Ministry requirement, not the Navy this time, for a general reconnaissance flying boat. Powered by Rolls Royce *Condor* engines and having a crew of five the wooden *Iris I* was rebuilt as *Iris II* with metal hull. The *Iris III* entered RAF service in 1930 and lasted until 1934 when they were replaced by the similar sized Blackburn *Perth*.

Between 1926 and 1932 while the *Iris* was being built Blackburn aircraft construction moved entirely to Brough where aircraft development continued throughout the difficult years of the Depression.

On the civil side the *B.2* was a small two seat biplane not unlike the de Havilland *Moth* which was its direct competitor. The *Baffin* continued the Naval tradition and was a torpedo carrier derived from the *Ripon*. Almost one hundred *Baffins* were built in 1932-33 including *Ripon* conversions but the type was short-lived, being replaced by the Blackburn *Shark* in 1933.

Built to an Admiralty specification for a torpedo-spotter-reconnaissance aircraft, the *Shark* was Blackburn's most successful type in production volume during the inter-war period. Ironically, the competitive Fairey *Swordfish* built to meet a similar need was also produced at Brough.

The mid-1930s were years of great expansion to meet the threat from Germany. At Brough the company name was changed to Blackburn Aircraft Limited on 2 April 1936. Additional capacity was created when a new factory was established at Dumbarton on the Clyde in cooperation between Blackburn Aircraft and William Denny & Bros. Ltd. Later versions of the *Shark* were built at Dumbarton which became the centre for Blackburn seaplane operations.

Blackburn became one of the few British companies to manufacture both airframes and engines. The Cirrus Hermes Engineering Co. Ltd., moved from Croydon to Brough in 1934. Two small four-cylinder in-line water cooled engines were produced; the *Cirrus Minor* of 90hp and the *Cirrus Major* of 150hp. Cirrus was taken over by Blackburn in 1940 to form Blackburn Engines and subsequently became part of Bristol Siddeley Engines in 1961.

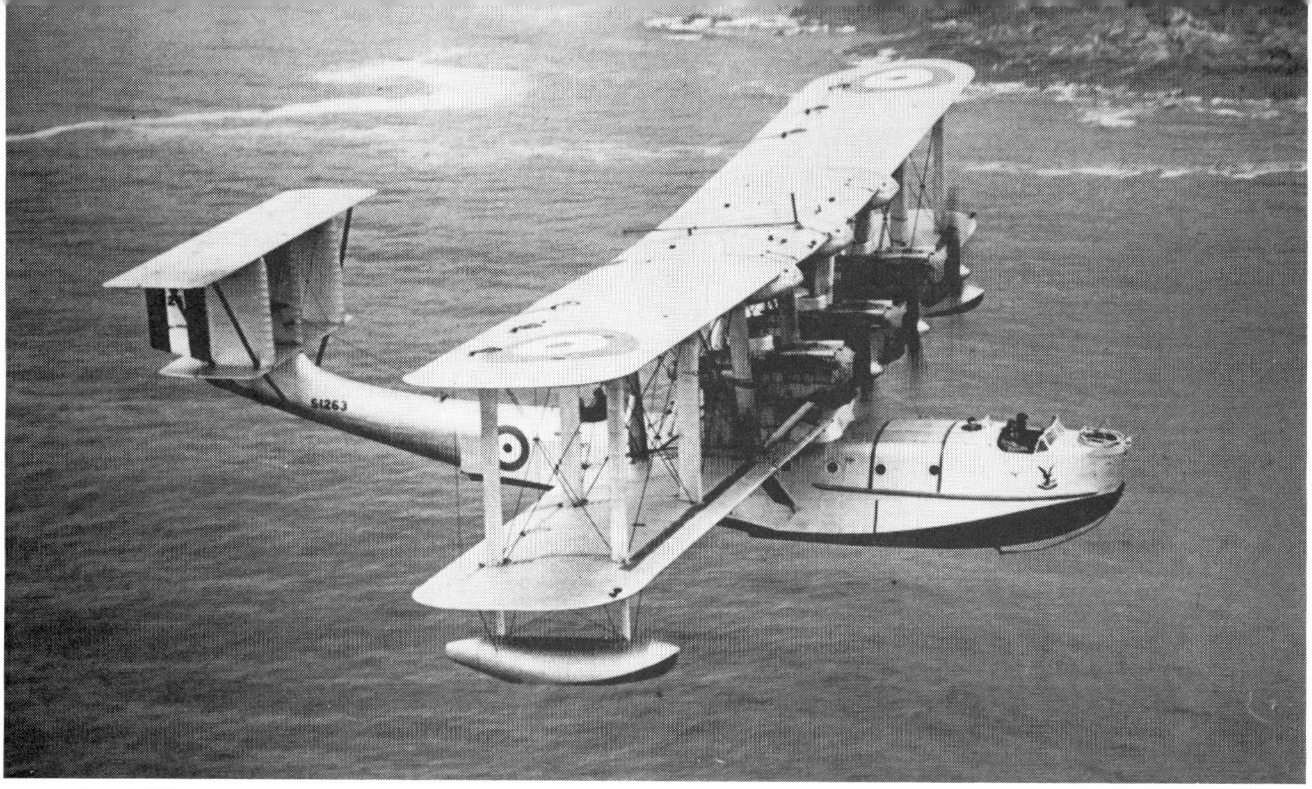
S1263

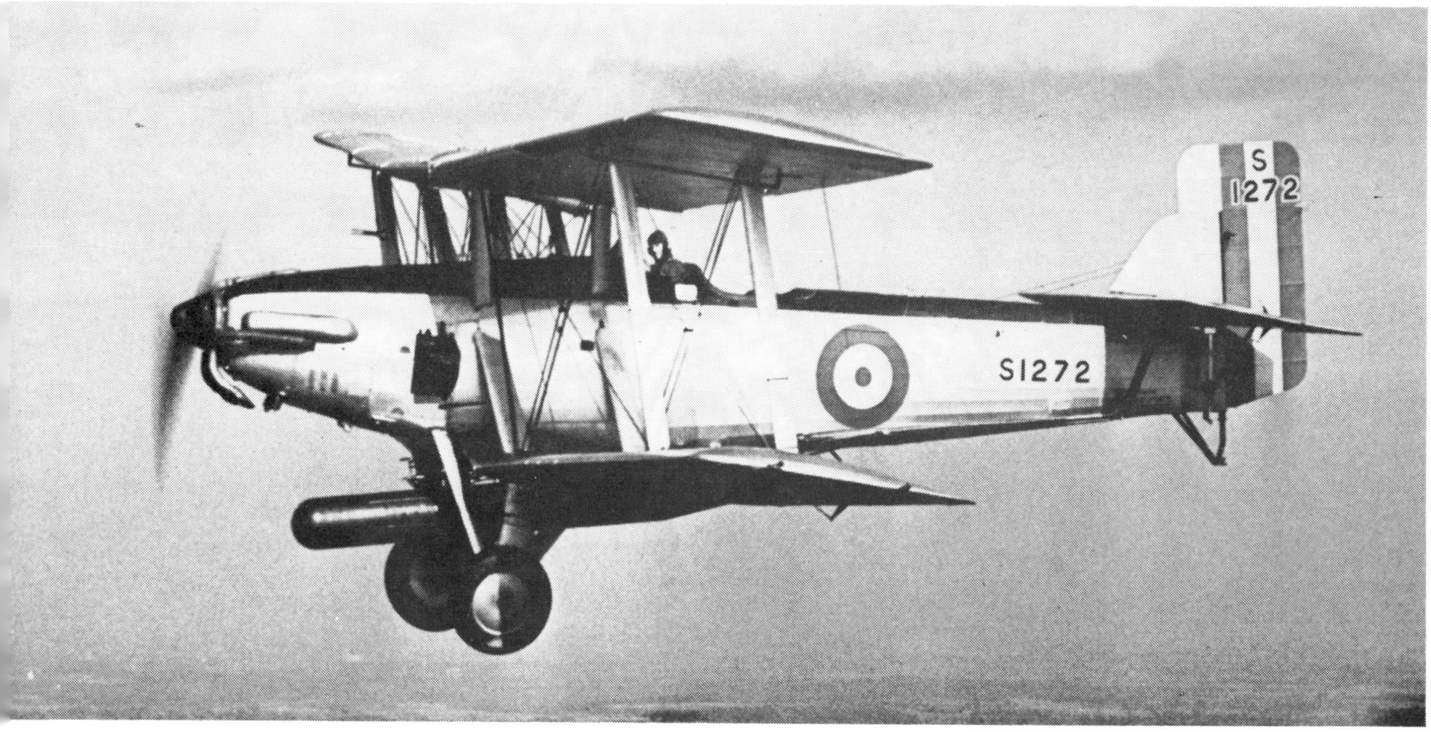
S
1272
S1272

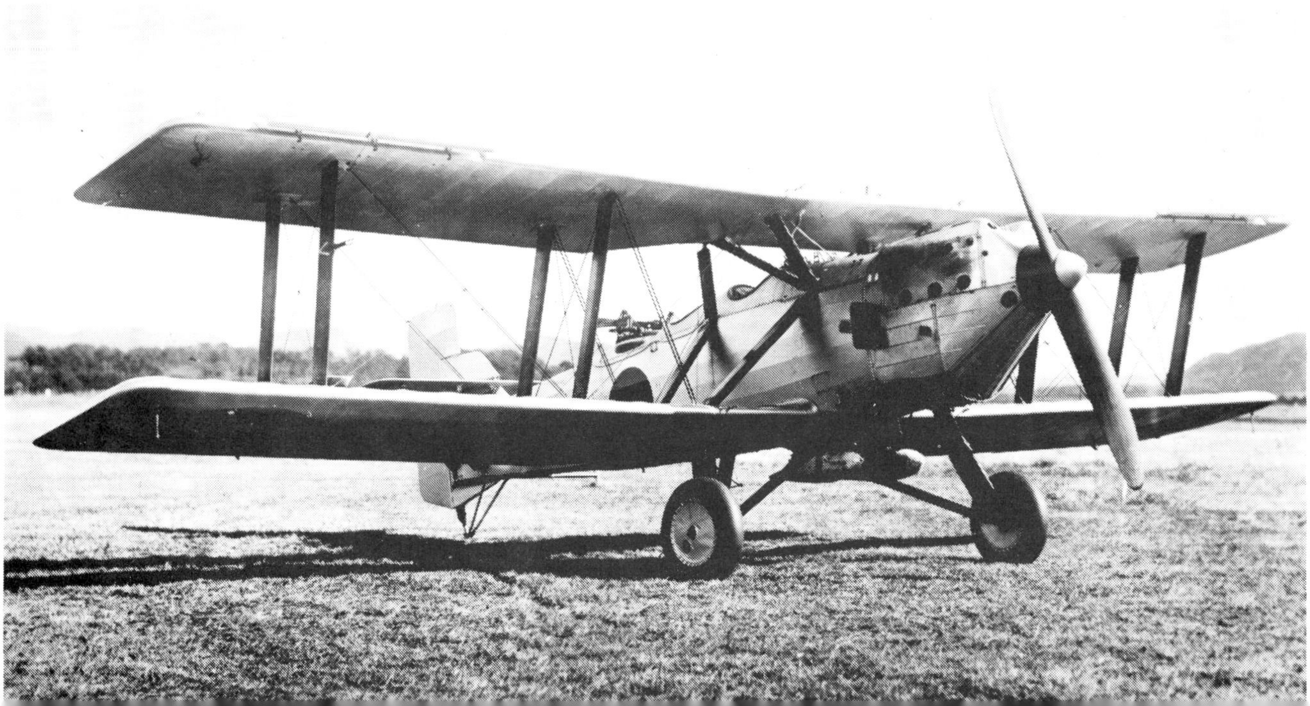

Left Upper:
BLACKBURN IRIS
Iris III (R1263) with three Rolls Royce Condor engines in service with No.209 squadron RAF. The Iris was the RAF's largest aircraft in the early 'thirties and appeared in five versions between 1926 and 1933.

Left Centre:
BLACKBURN RIPON
S1272 was a Ripon II, the production version of the Ripon torpedo carrier. Ninety-three Ripon II's with Napier Lion engines were built at Brough in 1927-30 and in addition, 25 Ripon IIF's were made in Finland in 1933.

Left Lower:
BLACKBURN 3MR4
Japanese-built 3MR4 with Mitsubishi-Hispano type 7 engine. One prototype, known as type T7B, was built at Brough and was followed by 204 Japanese 3MR4 aircraft. Note the Handley-Page auto slats on the upper wings.

Above:
BLACKBURN BAFFIN
K3589, the first production Baffin had a Bristol Pegasus I.M3 engine and was built in 1934. Twenty-nine Baffins were built from scratch and another sixty-eight machines were converted from Ripons.

Below:
BLACKBURN SHARK
K4352 was one of a batch of 16 production Shark I's built at Brough to equip No.820 squadron RAF. First flown in prototype form in 1933, the Shark I had an Armstrong-Siddeley Tiger IV engine. Later versions had Tiger VIC or Bristol Pegasus engines. Production totalled 251 aircraft including seven Pegasus engined Shark II's built in Vancouver by Boeing.

Above:
FAIREY SWORDFISH
Blackburn-built Fairey Swordfish II (HS158) from the first Brough batch. Production of the Bristol Pegasus engined Swordfish reached 1,700 at Brough; greater than any of Blackburn's own designs.

Left:
BLACKBURN SKUA
Production Skua II dive bomber (L2883) with dive brakes open. 311 Skua II's with Bristol Perseus engines were built in addition to the Bristol Mercury engined prototype which first flew in 1937.

Right Upper:
BLACKBURN ROC
L3084 was one of 136 Blackburn Rocs built under licence by Boulton & Paul at Wolverhampton. Shown here with Bristol Perseus engine, L3084 became a flying test bed at Filton.

Right Lower:
BLACKBURN BOTHA
Prototype Botha I (L6104) pictured at Brough at about the time of its first flight on 28 December 1938. Apart from the 380 Bothas built at Brough, another 200 were erected at Dumbarton. All had Bristol Perseus engines.

As the 1939 War seemed inevitable, official preference changed from biplanes to monoplanes in view of their superior speed and agility. Brough's first essay into all metal monoplanes created the *Skua* which first flew in 1937 and went into production as the Royal Navy's only special purpose dive bomber. Over three hundred *Skuas* were built but the similar *Roc* naval fighter was less successful.

Both the *Skua* and the *Roc* used the Bristol *Perseus* engine which was also chosen for the next Blackburn type to be built, the twin engined *Botha* general reconnaissance and torpedo bomber. The *Botha* went into production in 1938 but was found to be underpowered and with the adoption of the Bristol *Beaufort* as the torpedo bomber for RAF Coastal Command the *Botha* was issued to operational training units after 1940.

The *Firebrand*, Blackburn's best single engined aircraft of the 1939-45 war started life as a ship based interceptor fighter. Powered by a Napier *Sabre* engine the *Firebrand I* flew for the first time on 27 February 1942. To meet changing operational needs it evolved into a torpedo fighter with a Bristol *Centaurus* engine. *Firebrands* were produced until 1946 with later models being built as naval strike fighters.

To ensure that available resources were put to best use, all aircraft manufacturing came under the control of the Ministry of Aircraft Production in 1940. Under Lord Beaverbrook's direction orders were placed where a company's capacity and expertise was best suited to the Country's needs. Many commercial barriers were crossed as a result of the wartime effort.

As a consequence, Blackburn was given large orders for naval aircraft designed elsewhere. To meet the demands for increased production Blackburn had reopened the Olympia Works in Leeds and had also opened a new factory at an airfield at Sherburn-in-Elmet, Yorkshire.

Assembly of Fairey *Swordfish* biplanes was undertaken at Sherburn and 1,700 were turned out from December 1940 at a rate of sixty per month. Brough for its part assembled another Fairey type, the *Barracuda* monoplane torpedo dive bomber. *Barracuda* production was spread out by the Ministry among Fairey at Hayes and Heaton Chapel, Blackburn, Boulton Paul and Westland. The building of *Barracudas* followed on from the *Botha* and 700 of them left Brough in the early 1940s.

On the Clyde, the Dumbarton factory moved onto Short *Sunderlands* after *Botha* production expired and 250 of the Bristol *Pegasus* engined flying boats were delivered to the RAF.

With the ending of the war outstanding contracts for many military aircraft were cancelled. The Olympia and Sherburn works were closed. Brough took in general engineering work to supplement residual aircraft orders while Dumbarton turned to building prefabricated aluminium houses.

In 1948 Blackburn acquired the assets and business of General Aircraft Limited at Croydon and on 1 January 1949 a new company was formed called Blackburn & General Aircraft Limited. The Croydon works of G.A.L. was closed and all activity transferred to Brough.

At the time of the amalgamation G.A.L. had been working on a *Universal Freighter* powered by four Bristol *Hercules* engines. One aircraft was complete and another partially built. They were dismantled for road transport to Brough where they were rebuilt. The first GAL60 *(WF320)* was flown at Brough in June 1950 while the second was completed with more powerful Bristol *Centaurus* engines and flew in 1953.

From this second prototype was developed the *Beverley* heavy lift transport of which 47 were built at Brough for the RAF. The first production *Beverley (WZ889)* was first flown at Brough on 29 January 1955 and the last one *(XB259)* was delivered in May 1958.

Brough's first experience with jet propulsion was the Handley Page H.P.88 research aircraft built to flight test a reduced scale crescent wing for the Handley Page *Victor* programme.

Shortly before his death on 10 September 1955 Robert Blackburn had released vague details of a new naval strike fighter to be built at Brough. This was the N.A.39. *Buccaneer* which first flew as a prototype on 30 April 1958. Large orders were placed for *Buccaneers* to equip Fleet Air Arm squadrons of the Royal Navy. For flight testing the *Buccaneer*, Blackburn acquired an airfield at Holme on Spalding Moor. *Buccaneers* were towed on their wheels with wings folded from Brough Works to HOSM for flight. The first production *Buccaneer* flew from HOSM in 1962 and in the following year *Buccaneers* equipped the aircraft carriers HMS *Ark Royal* and HMS *Victorious.*

Robert Blackburn ruled the independent company that bore his name for fifty years but under the Government's amalgamation plans Blackburn & General Aircraft became part of the Hawker Siddeley Group late in 1959. Prior to this, the Company had been divided in early 1959 into four units; Blackburn Aircraft Limited, Blackburn Engineering Limited and Blackburn Electronics Limited. Blackburn Engines Limited already existed as a separate company.

Under Hawker Siddeley very little changed at Brough but more of this anon.

Left Upper:
BLACKBURN FIREBRAND

Firebrand T.F. Mk.IV (EK602) was the first of 104 such aircraft and first flew on 17 May 1945. Bristol Centaurus engines powered Firebrand III, IV and V while Firebrand I and II had Napier Sabres.

Left Lower:
GAL60

The prototype GAL60 Universal Freighter (WF320) which first flew at Brough on 20 June 1950 powered by four Bristol Hercules engines. This was the last aircraft to be built by General Aircraft Limited at Croydon before being taken over by Blackburn in 1949. An enlarged version with Bristol Centaurus engines became the Beverley.

Above:
BLACKBURN BUCCANEER

Pre-production N.A.39 Buccaneer (XK488) which first flew on 3 October 1958 powered by two de Havilland Gyron Junior turbojets. XK488 became a flying test bed at Filton until 1966 for testing the Gyron Juniors used in the Bristol 188 research aircraft. Later Buccaneers have Rolls Royce Spey engines. Buccaneers are being updated by British Aerospace and it is anticipated that they will remain in service well into the 1990s.

Below:
BROUGH

A recent photograph of the former Blackburn works (now BAe Brough) on the north shore of the Humber 10 miles from Hull.

Vickers Aviation

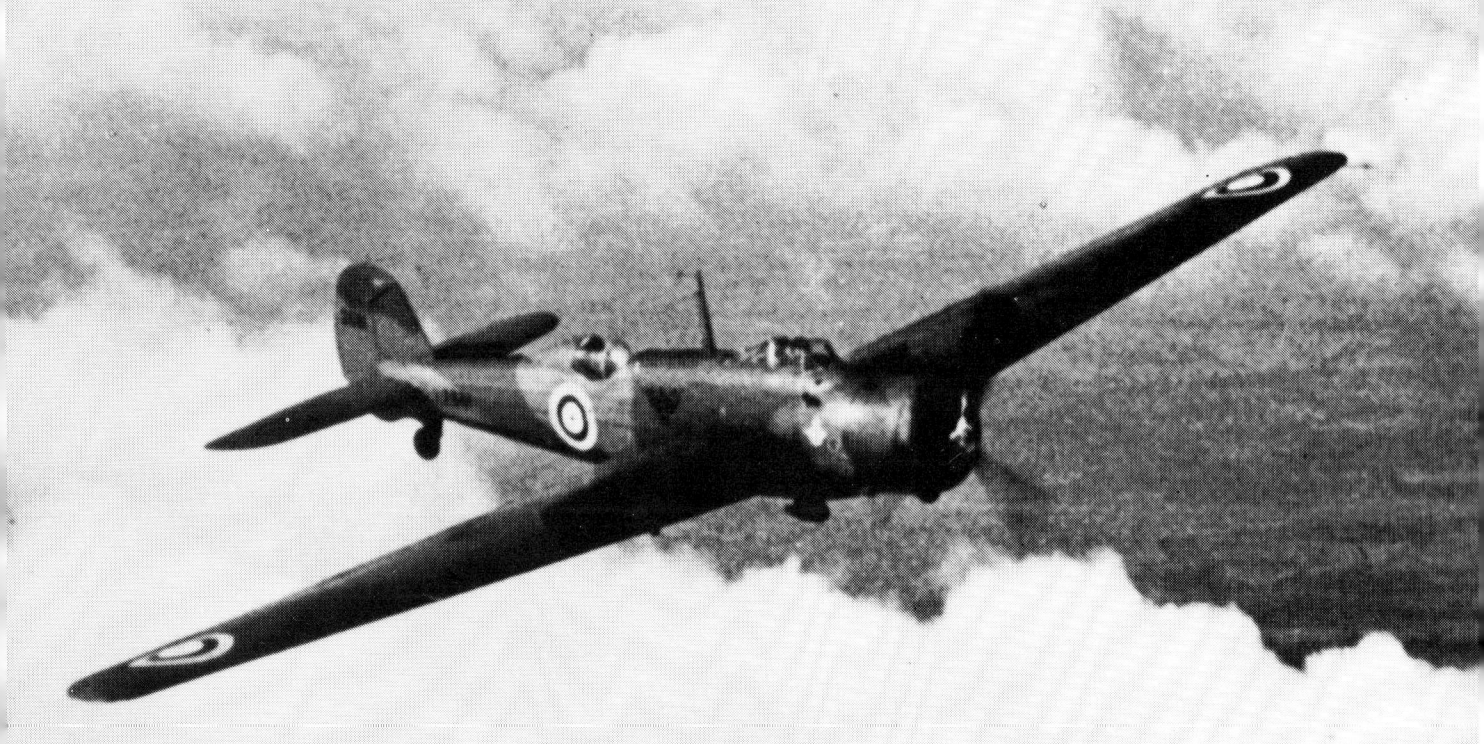

Mention of the name Vickers will conjure up in many people's minds pictures of the aircraft produced over a span of 72 years at Weybridge in Surrey. The Vickers enterprise does, however, go back to the early part of the nineteenth century and the Company has been at the forefront of innovation in many aspects of engineering. Aircraft was just one facet of Vickers activity and although given prominence here must be viewed as part of the overall picture of Vickers' achievement.

In 1829 Edward Vickers with the help of his brother William became involved in the establishment of Naylor, Hutchinson, Vickers & Company in Sheffield. This Company was new in title only for it continued part of the work of the firm of Naylor & Sanderson which had been dissolved. George Naylor, father-in-law of Edward Vickers, had by now retired and John Hutchinson was a silent partner so the enterprise was very much Vickers dominated from the outset.

The steel mill acquired by the Company at Mill Sands, Sheffield expanded steadily over the subsequent thirty years.

Besides the hot rolling of steel sections, Naylor Vickers did casting and forging for railways and other heavy industries.

Edward's son Tom became responsible for technical matters in 1851 while a younger son Albert took on financial affairs. As a result a new entity, Vickers, Sons & Company was incorporated in 1867.

Some score years later general steel making went into decline and Vickers was sustained by its special products such as guns and armour plate for battleships.

A trade depression in the early 1890s gave Vickers a difficult time. Yet another new company called Vickers, Sons & Maxim was formed in 1897 by the amalgamation of the Vickers interests with Maxim Nordenfelt Guns & Ammunition Co. Ltd., of Erith in Kent and Naval Construction & Armament Co. Ltd., at Barrow on Tyneside.

Starting with HMS *Vengeance*, which was completed in 1901, the new Company was able to supply ships complete with machinery, armour and armaments all from a single source. Vickers was always at the forefront of new ideas and in the early years of the twentieth century built American submarines under licence. In 1905 Conrad Clausse, a German national, took out a patent for *Duralumin*, an alloy of aluminium with a small amount of copper added which gave it the strength of mild steel, yet was very light. Shrewdly, Vickers obtained the U.K. manufacturing rights in 1910 and this was just one stage towards Vickers entry into aviation.

In the same year Vickers acquired the U.K. and Colonial manufacturing rights to the REP monoplane devised by the Frenchman Robert Esnault-Pelterie. The first Vickers *REP* was built at Erith in 1911 and seven were eventually flown powered by

Left:
VICKERS WELLESLEY
L2681 was one of three Pegasus- engined Wellesleys of the RAF Long Distance Flight which set up a World Distance Record of 7,157 miles on a flight from Egypt to Australia in November 1938.

Below:
VICKERS GUNBUS
This FB.5a Gunbus was built in 1915 for the Royal Navy Air Service and was numbered 870. It was later transferred to Royal Flying Corps duties and serialled 2345 as seen here. The FB5 was the first aircraft in the world built to mount a gun. *Vickers.*

either REP, Viale or Gnome engine. Actually, this type was not Vickers first experience in aviation because it had been loosely associated with the 'Captive Flying Machine' built by Hiram Maxim in 1894.

Under the guidance of Capt. Herbert F. Wood an Aviation Department was set up at Erith on 28 March 1911 and a flying school started at Brooklands in 1912. Farman type 'boxkites' were next purchased and a Vickers derivative soon appeared. This evolved through a series of fighter biplanes as the first true Vickers type; the *Gunbus* of 1914. Brooklands between Byfleet and Weybridge, was the focus of motor sport and the budding science of aviation. Consequently, Vickers set up shop in the old Itala Automobile Company's works next to the flying ground. Aircraft production commenced there in August 1915 with the B.E.2c biplane; a Royal Aircraft Factory (Farnborough) design for the Royal Flying Corps.

In the following year work commenced on another RAF type, the S.E.5a. Scout powered by a Hispano-Suiza engine. The first of these was delivered from Brooklands in June 1917 and over a thousand were built at a rate of thirty per week at the peak.

Vickers' past experience well suited it to solving new problems and at about the time of the start of the Great War new ideas were forthcoming. The reason for the rear engine arrangement of the *Gunbus* was to give an unrestricted field of fire for the guns. On forward mouted engines there was the risk of bullets shooting away the propeller. George H. Challenger, who had designed the Bristol 'Boxkite' before joining Vickers, invented a gun synchronising gear in 1915 which was first tested in a Sopwith *'1 ½ Strutter'*.

George Challenger was the first man in the Aircraft Department to be bestowed with the title of Chief Designer, a position which he held until 1914. His team included other ex-Bristol men notably Archibald Low, a design draughtsman, and Leslie Macdonald, a pilot. When the aircraft design office transferred to Crayford in August 1914 (later in WW1 it was at Imperial Court, Knightsbridge), R.K. Pierson became Chief Designer, a post he retained until 1945.

From single engined fighters Rex Pierson moved on to multi-engined bombers and the *Vimy* was an important milestone in the evolution of Vickers aircraft. First flown in the hands of Capt. Gordon Bell on 30 November 1917 at Joyce Green in Kent, it was the culmination of experience on several twin-engined experimental machines. The signing of the Armistice almost a year later prevented the *Vimy* from bombing German targets but it was to be the first of several successful Vickers twin engined bomber/transports.

The first of these was the *Vimy Commercial* which featured a bulbous fuselage of wooden construction married to *Vimy* wings and tail. Like the *Vimy II* it had a pair of Rolls Royce *Eagle* engines and was followed by the *Vernon*, a bomber-transport and troop carrier. Later *Vernons* had Napier *Lion* engines and the experience with this engine led to the *Virginia* bomber of which 122 were built from 1922 onwards. The installation of the Bristol *Jupiter* and *Pegasus* engines extended the Service life of the *Virginia* well into the 1930s.

Further development of the Vimy led to the *Victoria* troop carrier having *Lion* engines. The wooden prototype, *J6860*, flew on 22 August 1922 and by 1928 the all metal version, *Victoria IV*, was underway.

Conversions of the *Lion* engined *Victoria V's* to Bristol *Pegasus* engines created the *Victoria VI* which had the potential of higher gross weight. To differentiate the improved airframe a new name *Valentia* was adopted in 1934 for subsequent machines, mostly conversions. At this stage a civil

Below:
VIMY COMMERCIAL
G-EASI, christened "City of London" by its operator S. Instone & Company inaugurated the Croydon-Paris passenger service in 1921. Developed from the Vimy bomber of WW1, the Vimy Commercial is regarded by many people as the world's first airliner.

version named *Vanguard* was proposed but only one was built.

While Weybridge had been involved in twin engined transports for the RAF the single engined machine had not been forsaken. Under Pierson, the design office was prolific in ideas between the early 'twenties and the start of the rearmament drive in the 1930s. There is insufficient space here to describe them all.

The main prospect in the 1920s was the Bristol *Fighter* replacement and for this role Vickers offered the *Vixen* in 1923. Using the well proven Napier *Lion* it did not please the British authorities but was sold to Chile and a variant called *Valparaiso* was sold to Portugal. Later versions were aimed at the army cooperation role and the *Venture* was followed by the *Vivid* and *Valiant* which were metal framed.

Under the Defence of the Realm Act an Excess Profits Duty was levied on all companies and Vickers did not escape the tax that bankrupted many others. At the end of the war a Peace Products Committee was set up within Vickers to find new sources of revenue. The Company diversified but perhaps too much and this action was to have repercussions in later years.

The Nation was disrupted by the General Strike in 1926 but besides this the industrial slump had caused a crisis in Vickers. Armstrong-Whitworth was Vickers biggest rival for armament and shipbuilding orders and this Company too was in difficulty in the mid-1920s.

On 31 October 1927 the industrial giants merged to form Vickers-Armstrongs Limited. In early 1928 the aircraft side of the business was renamed Vickers (Aviation) Limited as a wholly owned subsidiary of Vickers-Armstrongs. Here it should be explained that Vickers did not gain control of Armstrong Whitworth Aircraft which continued as a subsidiary of Armstrong Siddeley Development Company. This was not the end of the merger trail for Vickers took over the Supermarine Aviation Works in 1928.

Supermarine was formed in 1913 by Noel Pemberton-Billing at Woolston on the river Itchen near Southampton. The name was an invented one for seaplanes which flew over water as opposed to submarines that travelled under it. After the merger it became The Supermarine Aviation Works (Vickers) Limited in 1931.

Supermarine had one particularly valuable asset in R.J. Mitchell, a railway engineer who had joined the Company in 1917. He is best remembered for his *Spitfire* which Mutt Summers took up for the first time on 5 March 1936, shortly before the untimely death of its creator. Although Mitchell takes the initial credit for the design, the Supermarine team worked hard until the end of the war to get even more out of what was already a truly remarkable fighter.

Previously on 1 January 1930, another outstanding designer, B.N. Wallis had been appointed Chief Designer (Structures) at Weybridge following the cancellation of the *R.100* airship. Barnes Wallis' involvement in airships went back to before the First World War when, as an apprentice, he had worked under H.B. Pratt on Airship *No.9* at Walney Island near Barrow. His Majesty's Airship *No.9* was the second rigid airship to be built by Vickers.

Wallis' had a leading influence in the design of the later *R.80* airship in which he rejected the traditional sausage shape in preference to a curved and more aerodynamic profile.

In the 1920s it was proposed that the airship could be the ideal vehicle to service the long distance 'Empire' routes to India and Australia. The Airship Guarantee Company was set up in 1923 to develop a large airship similar to the *R.80* and work commenced at Howden, Yorkshire in 1926. *R.100* first flew on 16 December 1928 and successfully flew to Canada and back again. Unfortunately, when the Government

Above:
VICKERS VIRGINIA
J7717 was originally built as a Virginia VI with two Napier Lion II engines. It was later given more powerful Lion power plants and became Virginia X. *Vickers.*

Below:
VICKERS VALENTIA
Vickers type 264 Valentia with two Bristol Pegasus engines. *Vickers.*

Above:
VICKERS VINCENT
K4105, the first Vildebeest Mk.III as converted to Vincent with Bristol Pegasus II.M3 engine. *Vickers.*

Below:
VICKERS VIASTRA
Type 259 Viastra X, G-ACCC powered by two 650hp Bristol Pegasus II.L3 engines. Several Viastras were flown and were part of a family of all metal airframes built to exploit the patents of Michel Wibault. *B.E.D.*

sponsored *R.101* crashed at Beauvais, France on 6 October 1930 with great loss of life, the private venture *R.100* became an acute embarrassment to the Labour Government and it was ordered to be scrapped.

At Weybridge, Barnes Wallis was given free expression of his ideas and in particular his form of light alloy geodetic or basket weave construction. An order was received from the Air Ministry for a torpedo bomber biplane and in addition to this Vickers proceeded to build a monoplane version as a private venture. In official trials the monoplane had superior performance and was adopted for production as the *Wellesley*. Powered by a Bristol *Pegasus* engine the *Wellesley* prototype was flown on 19 June 1935 and 176 were in service by March 1938.

Vickers Aviation and Supermarine continued to operate as separate concerns aloof from the Vickers-Armstrong parent company until 1938 when closer control was established. Weybridge assumed the title of the main company while Woolston became known as Vickers-Armstrongs Limited (Supermarine Division). Due to the impending war the product lines remained unchanged.

These two firms had a very important role in the run up to the 1939-45 war with production of the *Wellington* and *Spitfire* respectively. Under the 'Shadow' scheme other factories were built by the Government at Hawarden near Chester, and Squires Gate, Blackpool to build *Wellingtons* and Castle Bromwich in the Midlands for *Spitfires;* all three being managed by Vickers.

Built to Air Ministry specification B9/32 the prototype *Wellington, K4049,* made its maiden flight from Weybridge on 15 June 1936. *K4049* had two Bristol *Pegasus* engines driving de Havilland-Hamilton two speed airscrews and this was standardised for early production *Wellingtons.*

Variations to the *Wellington* bomber involved alternative engine types to ensure continuity of supply and maximum performance. *Wellington Mk.II's* used two Rolls Royce *Merlins* and flew in this form on 3 March 1939. Bristol *Hercules* engines in the *Wellington Mk.III,* which flew on 19 May 1939, gave superior performance to the *Pegasus* powered Mk.I's. Pratt & Whitney *Twin Wasp* engines were adopted for the *Wellington Mk.IV* and improvements to the airframe through the use of high strength light alloys gave rise to the *Wellington X*.

'Wimpies' had an active life and saw many variants and modifications for special duties. They were the mainstay of RAF Bomber Command in the early years of the war until replaced by the 'heavies' like the *Lancaster, Stirling* and *Halifax.*

The last Weybridge *Wellington*, a Mk.XIV, was out-shopped in September 1943 and was the 2,515th of the breed to be built in Surrey.

By this date Vickers had started work on its successor, the geodetic *Warwick* of which 845 were built at Weybridge up to 1945. The four engined *Windsor* built at the same time was not so lucky and only three prototypes were built before the project was cancelled at the end of the war with Japan.

Before peace was restored, a committee chaired by Lord Brabazon of Tara had drafted specifications for future British airliners. Vickers offered the *Viscount* but as an interim measure it was proposed to build a small *Vickers Commercial One* based on wartime products.

For the VC1 the *Wellington* was taken as a basis but a new all metal fuselage was created and a *Warwick* tail was used. Powered by two Bristol *Hercules 100* engines the prototype VC1 *Viking* was first flown at Wisley by Mutt Summers on 22 June 1945. From the twentieth aircraft the *Viking* had an all metal structure and entered service with BEA, Air India, Air Lingus, South African Airways and others.

Military versions of the *Viking* were produced for the Ministry of Supply including the *Valetta* paratroop transport with large doors in the fuselage side aft of the wings. The *Varsity* was a multi-engined bombing trainer having a bomb aimer's bulge under the fuselage.

By 1948 the Vickers VC2 to meet the Brabazon IIB specification was a reality when the prototype *Viscount 630* was flown from Wisley on 16 July by Mutt Summers. With four of the new Rolls Royce *Dart* turboprops, the *Viscount* was destined to be Vickers big seller in the 1950s. Originally intended for BEA it was also sold to Trans-Canada Airlines, forerunner of Air Canada, as well as the United States and many smaller operators around the world.

Above:
SUPERMARINE SPITFIRE
RW396, a Spitfire L.F.XVI, was built at Castle Bromwich shadow factory at the end of the 1939-45 War. The Spitfire was the only aircraft to stay in production for the full duration of the War. Early examples had Rolls Royce Merlin V-12 engines but later marks had more powerful Griffons. MOD.

Below:
VICKERS WELLINGTON
Rex Pierson and Barnes Wallis' geodetic bomber affectionately known as the "Wimpy" was the RAF's front line bomber in the early years of the 1939 war. P9249, a Wellington I.C., had two Bristol Pegasus XVIII engines.

Production commenced at Weybridge but nearly two-thirds of the final total were erected at the Vickers works at Hurn Airport. In 1941, an RAF station had been established at Hurn, Bournemouth and this became a civil airport under Ministry control in 1946. Vickers took over the hangarage at Hurn from BOAC in 1951 for flight testing of the *Valiant*. Hurn was subsequently developed as a works for final assembly of *Varsity* and *Viscount* aircraft using details and sub-assemblies sent from Weybridge.

True to form, Weybridge did not forsake the heavy bomber and received orders for the *Valiant* strategic jet bomber as its contribution to the nuclear strike force. The *Valiant* was less sophisticated than the Avro *Vulcan* or Handley Page *Victor* but on this account it scored because it could be put into service relatively quickly. The prototype, *WB210*, was test flown on 12 January 1952 but crashed due to an engine fire. A second prototype flew on 11 April 1952 and the *Valiant* entered RAF service in 1955.

On 1 January 1955 a corporate reorganisation took place within Vickers. A new subsidiary company was formed called Vickers Armstrongs (Aircraft) Limited to coordinate all Vickers and Supermarine activity under one management.

Above:
VICKERS WARWICK
Second prototype Warwick bomber (L9704) probably with Pratt & Whitney engines although it also flew with Bristol Centaurus units. Conceived as a replacement for the Wellington, Warwicks were actually deployed by RAF Coastal Command for general reconnaissance and anti-submarine work. *Vickers.*

Below:
VICKERS VIKING
VP-YEW, a Viking 1B, sold to Central African Airways in 1946. Powered by two Bristol Hercules engines and using major components of the Wellington and Warwick, the VC1 Viking was Britain's first post-War airliner.

Above:
VICKERS VARSITY
Varsity prototype, VX828, with Bristol Hercules engines which first flew in July 1949. Similar to the Viking but having a tricycle undercarriage, the Varsity was adopted by the RAF for navigation, bomb-aiming and advanced pilot training. *B.E.D.*

Below:
SUPERMARINE SCIMITAR
XD212, the first production Scimitar which attained first flight in January 1957. The R.R. Avon-engined Scimitar was the Fleet Air Arm's first supersonic jet aircraft and was capable of ground attack and could carry tactical nuclear weapons. Scimitars were built at South Marston near Swindon in Wiltshire. *Vickers.*

After the success of the *Viscount*, Vickers devised another airliner based on the same formula; a medium range machine powered by four turboprops. Unfortunately Weybridge was less successful with this one partly due to the passenger appeal of the pure jets and partly because the Rolls Royce *Tyne* engines, which had received Government support to the detriment of the Bristol *Orion*, were troublesome at first.

With the obvious exception of the *Valiant*, most of Vickers turbojet experience since the war had been at Woolston where orders had been received from the Navy for the *Attacker*, first flown on 27 July 1946, and the *Scimitar* which flew on 19 January 1956. The *Swift*, intended as a *Meteor* replacement had to settle for second best to the Hawker *Hunter*.

On 6 April 1948, Weybridge took the credit for the first turbojet civil transport in the world when a *Viking (VX856/G-AJPH)* belonging to the Ministry of Supply was flown experimentally with two Rolls Royce *Nene* engines in underwing pods in place of the standard *Hercules* piston engines. A similar experiment took place with the Rolls Royce *Tay* powered *Viscount* test bed *(VX217/G-AHRG)* which first flew on 15 March 1950.

A start was made at Weybridge in 1951 on the V1000 military transport and a civil version to be called VC7. A prototype of the V1000 was partly built but was cancelled in 1955 and this left Vickers in the unenviable position of having no large transport aircraft to offer.

A BOAC requirement for a long range jet airliner capable of using some of the more difficult airports resulted in Vickers last type as an independent company.

Developed initially for BOAC routes to Africa and the Far East the 135 passenger VC10 evolved as a 'universal' airliner which could also be used on BOAC's North Atlantic service. New buildings were erected at Weybridge for VC10 production and two euphemistically named the 'Cathedral' and 'Vatican' were used for final assembly.

Above:
VICKERS VISCOUNT
ZK-BRD "City of Wellington" was one of four Viscount 807's sold to New Zealand. The Rolls Royce Dart- engined Viscount was the world's first turboprop airliner to go into revenue service and were bought by sixty operators in forty countries. Most of them were erected at Hurn Airport in Hampshire.

Below:
VICKERS VALIANT
Valiant B.(K).1 serial XD823 in white 'anti-flash' finish and equipped for in flight refuelling. The Rolls Royce Avon powered Valiant was the RAF's first 'V'-bomber and earned the dubious distinction of dropping the first British hydrogen bomb on Christmas Island in the Pacific. A total of 104 Valiants was built at Weybridge.

Above:
VICKERS VC10

The second production VC10, G-ARVB, built for BOAC in 1962. Powered by four Rolls Royce Conway turbojets they were first used on the London-Africa routes. Most VC10 s, the last true Vickers aircraft, were built after Vickers Armstrongs (Aircraft) had become a part of British Aircraft Corporation; the last (Super) VC10 being built in 1970.

Left below:
VICKERS VANGUARD

G-APEB, the second production Vanguard which first flew on 23 July 1959. Designed expressly for BEA to replace the Viscount, the R.R. Tyne-engined Vanguard also found favour with Trans-Canada Airlines but did not achieve its full potential due to the emergence of turbojet powered airliners.

Below:
WEYBRIDGE WORKS

The former Vickers' works between Byfleet and Weybridge photographed in 1968. Built within the old Brooklands motor racing circuit, the airfield was the testing ground of many aircraft firms and individual aviators.

G-ARTA, the prototype VC10 flew from Weybridge to the Wisley flight test centre on 29 June 1962 by which time Vickers was a parent of British Aircraft Corporation. VC10 s went into service to Lagos in April 1964.

Having proved itself very attractive for North Atlantic crossings full advantage was taken of the VC10s potential by 'stretching' the fuselage by 13ft and increasing the gross weight. The first of these 163 passenger Super VC10 s flew on 7 May 1964 and went into service on the London - New York run on 1 April 1965.

Attracted by the VC10, British United Airways specified a side loading door for its two aircraft and this resulted in orders for similar cargo/passenger VC10 aircraft for Ghana Airways and the RAF, and five Super VC10s for East African Airways.

Production was to total thirty-one VC10 including fourteen for RAF Air Support Command and twenty-two Super VC10.

Delivery of the last Super VC10 for East African Airways *(5H-MOG)* brought Vickers production to a close in 1970 but Weybridge was by then deeply involved with the BAC *One-Eleven* and *Concorde*.

Vickers was an armaments manufacturer which took to aircraft building when the Company was already fifty years old. Together with the companies that were absorbed along the way Vickers built aircraft until it finally relinquished its interests at Weybridge to British Aircraft Corporation in 1960 and in turn to British Aerospace in 1977. Thereafter it reverted to being a wide ranging engineering company.

HAWKER

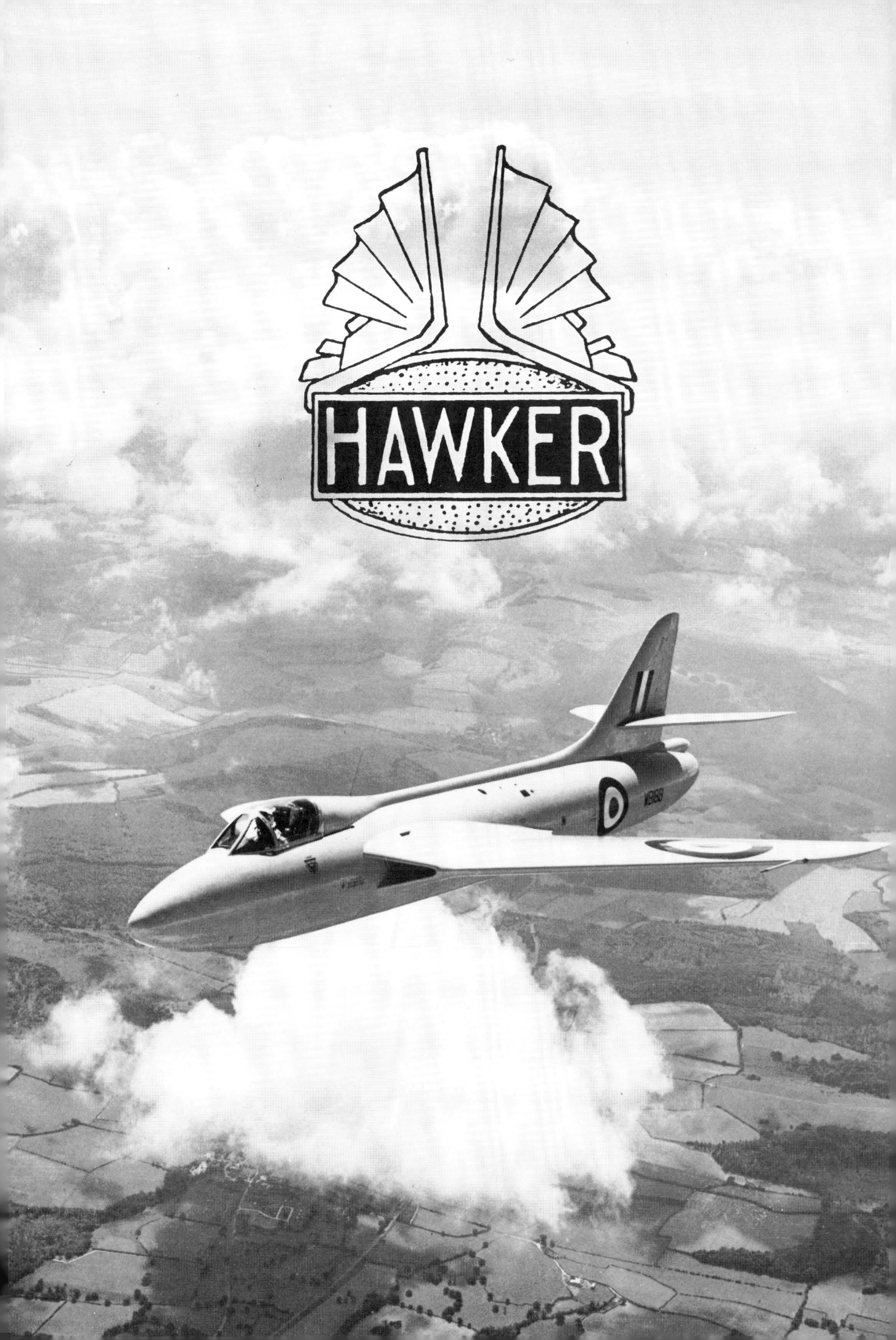

HAWKER

Hawker aircraft have become world famous but had it not been for one of those curious twists of fate it would be the name Sopwith that would now be prominent.

Thomas Octave Murdoch Sopwith, the son of a wealthy civil engineer, was born on 18 January 1888. He was one of those fortunate young men who grew up in the late-Victorian and Edwardian eras to whom money was no object and the world was there for the taking.

He had a passion for fast boats and part-owned a motor boat with a friend, Bill Eyre. Another member of his circle of friends was Fred Sigrist who was mechanically adept and spent much of his life below decks as a result.

Attracted by the novel sport of flying he soon bought himself a Howard Wright monoplane and taught himself to fly. This was just the first of several machines that he owned for pleasure but his first 'job' in aviation was as test pilot for the early biplanes designed by W.O. Manning and built at Battersea by Coventry Ordnance Works in 1912.

The story of the opening years of the Sopwith dynasty is sinuous and involved but suffice it to say that a move was made towards making aeroplanes rather than just playing with them. Sopwith Aviation Company was formed and a disused roller skating rink in Canbury Park Road, Kingston-upon-Thames was acquired as working premises. A Sopwith flying school was also established at Brooklands.

At the Aero Show held at Olympia, London in February 1913 a Sopwith *Three Seater* was displayed, this being the first positive sign of Sopwith's intention to be an aeroplane constructor.

1913 also saw the introduction of the *Tabloid* scout which was to see service in the opening years of the Great War. It presaged a long line of biplanes to be built by Sopwith Aviation and subcontractors.

Mention of subcontracting brings to mind the '1½ Strutter' built in 1915 which was the first Sopwith machine to be built in large quantity by outside firms.

In its short life of only eight years Sopwith turned out about fifty different types of aeroplane for the Navy and Royal Flying Corps. Some were only prototypes while others like the *Pup, Camel, Dolphin, Snipe* and *Salamander* were built in hundreds.

At the end of the war the Government imposed an Excess Profits Duty and, unwilling to continue in business with a heavy financial burden, Sopwith Aviation Company went into voluntary liquidation in 1920.

Above:
H.G. HAWKER 1889-1921
Pioneer aviator Harry George Hawker with an early Sopwith Three Seater biplane. Born in Australia, he was a leading influence in the design of Sopwith machines.

Left:
T.O.M. SOPWITH 1888-
Thomas Sopwith with his Howard Wright biplane circa 1911. The founder of the Sopwith Aviation Company 1912-1920 he was the leading figure in the Hawker companies from 1920 until his retirement. He was knighted in 1953.

Opposite Page:
HAWKER HUNTER
First prototype Hawker P.1067 Hunter, WB188, with Rolls Royce engine as first flown on 20 July 1951.

After the demise of the Sopwith enterprise a new entity, H.G. Hawker Engineering Company Limited, was incorporated on 20 November 1920. Thomas Sopwith was Chairman and other 'Sopwith' men who became directors were Fred Sigrist, Major Eyre, F. Bennett and of course Harry Hawker after whom the Company was named. The registered office of Hawker Engineering was at the old Sopwith factory at Canbury Park Road, Kingston-upon-Thames. With a dearth of aircraft projects after the 'war to end all wars', Hawker turned to general engineering and built cars and motorcycles among other products.

It was not long, however, before the directors were overcome by the yearning to get back into the air again. Initially some work was obtained in reconditioning de Havilland D.H.9A's, Sopwith *Snipes* and Sopwith *Camels* which was undertaken at the Sopwith factory in Canbury Park Road. Some engine overhaul work was also forthcoming and was done at Brooklands.

Unfortunately, the new company suffered a severe setback when Harry Hawker was killed when the Gloster *Nieuport* he was flying crashed on 12 July 1921. He had been unwell for some time and died of natural causes aggravated by test flying.

In 1922 Capt. B. Thompson was appointed chief designer and created the *Duiker* high wing monoplane which was flown, albeit unsuccessfully, in 1923 first with an Armstrong Siddeley *Jaguar* and later a Bristol *Jupiter IV* engine. His other design, a biplane, was the *Woodcock* night fighter having a *Jaguar* radial engine. Neither of these machines went into production and Thompson subsequently left Hawker.

W.G. Carter, later to become one of Britain's top aircraft designers, joined Hawker in 1924. Carter devised a new set of wings for the *Woodcock* and with a Bristol *Jupiter IV* engine it went into production in 1924 as the *Woodcock II*; Hawker's first production machine. After only a year at Kingston, Carter left to join the Gloucestershire Aircraft Company (later Gloster Aircraft Company Limited) where he was responsible for a succession of types including Britain's first operational 'jet' aircraft.

A young design draughtsman from the defunct Martinsyde company joined Hawker and had become George Carter's assistant. When his boss left to seek greener pastures Sydney Camm was appointed chief designer. Camm's first task was to adapt the *Woodcock* to the needs of the Danish Government which ordered three, named *Danecock*. Twelve more were built under licence at the Danish Royal Naval Dockyard. His first original design was the *Cygnet* ultra-light biplane built for the Light Aeroplane Competition at Lympne in 1924.

Hawker's largest biplane and the last to be wooden framed, the *Horsley* bomber entered service with the RAF in 1927. Wood gave way to metal framing starting with the *Heron* lightweight fighter built as a private venture in 1926 to test the method of tubular framing patented by Hawker.

The first flight of the prototype *Hart* two seat day bomber in June 1928 signalled the start of Hawker's involvement as suppliers of front line light bombers and fighters for the RAF. Development of the *Hart* into a family of types to suit different applications enabled aircraft to be produced quickly at minimum cost and thus maximum profit.

In 1928 the Ham Works built in WW1 and at that time surplus to requirements was leased to Leyland Motors.

Over one thousand *Hart* bombers of various configurations were built by Hawker, Gloster and Vickers. From it was derived the single seat *Fury* interceptor fighter which retained the Rolls Royce *Kestrel* engine as used in its predecessor giving it a very sleek appearance and speed to match its looks. Deliveries to RAF units commenced in 1931 and the type was exported to Norway, Jugoslavia, Persia (Iran) and Spain.

Left Upper:
SOPWITH DOLPHIN
Sopwith type 5.F.1. Dolphin (C3786) with twin Lewis guns and powered by a 200hp Hispano-Suiza engine. Production of the Dolphin at Kingston and by two sub-contractors between 1917 and 1919 totalled 1,532.

Left Lower:
SOPWITH CAMEL
Sopwith type 2.F.1 Camel with 150hp A.R.I. engine. Armed with a wing mounted Lewis gun and a Vickers gun above the fuselage N6635, pictured here at Brooklands, was built at Kingston for the Royal Naval Air Service in 1917.

Above:
HAWKER WOODCOCK II
The Bristol Jupiter IV engined Woodcock II was the first Hawker aircraft to go into production and to enter RAF service. Sixty-three Woodcock II's were built at Kingston.

Below:
HAWKER HORSLEY
S1452 with Rolls Royce Condor III engine. The Horsley bomber was the largest Hawker piston-engined aircraft to be built and the first to be produced in large quantity. 121 Horsleys were built at Kingston.

For the Royal Navy the *Nimrod,* evolved by way of the private venture *Hoopoe,* was a fleet interceptor for which duty it was strengthened for catapult launching and 'hooked' for deck landing. One *Nimrod* was sold to Japan which was acquiring as many of the best types as it could. Like Bristol and others, Hawker was caught out in this respect because the Japanese simply copied the design without paying royalties.

Known originally as the *Hart Fighter,* the *Demon* two seat fighter went into production in 1932 and was later supplemented by the *Osprey* fleet spotter which was built in quantity in the following year. To improve its versatility, seaplane versions of the *Osprey* were built but sales did not match those of the wheeled variety.

Built as a replacement for the Armstrong-Whitworth *Atlas,* the *Audax* army cooperation biplane flew on 29 December 1931. Most of the 618 to be built had *Kestrel* engines but machines sold to Iraq had Bristol *Pegasus* radial engines while the *Persian Audax* used, alternatively, *Pegasus* or Pratt & Whitney engines. To meet sales commitments, *Audaxes* were built by Gloster, Bristol, Avro and Westland besides the genuine Kingston product.

There were several members of the *Hart* family that did not proceed beyond the prototype stage including the *PV3,* aimed to replace the Gloster Gauntlet, and the *PV4* general purpose biplane.

The last important link in the chain was the *Hind* general purpose bomber which flew in September 1934 and was to total 582 aircraft, all built at Kingston.

Expansion of the aircraft manufacturing activity and the commercial success of the *Hart* series set the stage for corporate changes in the 1930s. On 18 May 1933 Hawker Aircraft Limited was created as a private limited company to take over the assets and business of H.G. Hawker Engineering Co. Ltd., which then ceased to exist.

Many Hawker aircraft had to be subcontracted in the early 1930s due to lack of capacity at Kingston. To rectify this situation Hawker Aircraft Limited acquired one hundred percent control of Gloster Aircraft Limited which had a factory at Hucclecote but was in dire straits through lack of work although Gloster had acquired an RAF order for *Gauntlet* fighters. With the expansion of the industry generally to meet the threat of Hitler's new Luftwaffe the time was ripe for take-over bids.

With the Government's blessing a new entity, Hawker Siddeley Aircraft Company Limited was formed on 11 July 1935 to amalgamate half of the assets of Hawker Aircraft Limited with the total assets of Armstrong Siddeley Development Company Limited which in turn controlled A.V. Roe & Company Limited and Sir W.G. Armstrong Whitworth Aircraft Limited.

Left:
HAWKER HART
Hart J9941 with Rolls Royce Kestrel engine is now preserved. First flown in 1928, the Hart day bomber and its derivatives sustained the Hawker works in the 1930s. A total of 1030 Harts was built with many variations.

Left Lower:
HAWKER FURY
Portuguese Fury No.50, the first of three sold to the Portuguese Army Air Force, pictured at Brooklands probably at the time of its first flight in May 1934. The Kestrel engined Fury was a single-seat interceptor fighter based on the Hart. Nearly 300 were built at Kingston and by General Aircraft Limited.

Right Upper:
HAWKER AUDAX
Kingston built Audax K2012. The Audax was a two-seat army cooperation variant of the Hart and the prototype first flew in 1931. Over 600 were built by Hawkers, Avro, Bristol, Gloster and Westland.

Right Centre:
HAWKER DEMON
A1-1 illustrated was the first of 16 Demon I's sold to the Royal Australian Air Force in 1935. A two-seat version of the Hart, about 300 Demons were built and A1-1 was probably built by Boulton and Paul at Norwich.

Right Lower:
HAWKER HIND
K6689 was one of a batch of 244 Hinds built for the RAF in 1935. Yet another variant of the Hart family, the Hind was a two-seat general purpose bomber and trainer with a Rolls Royce Kestrel engine.

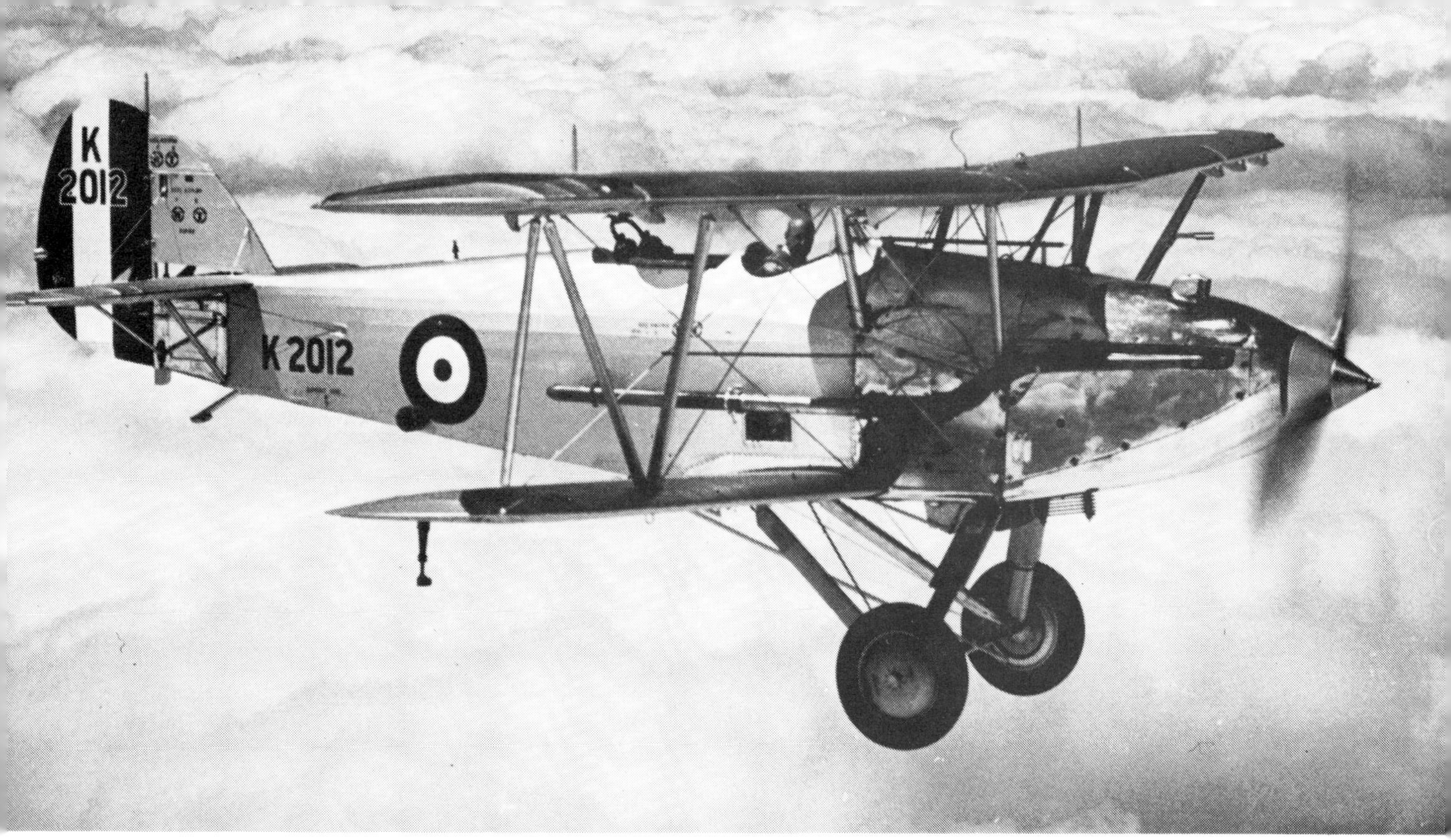
K
2012
K2012

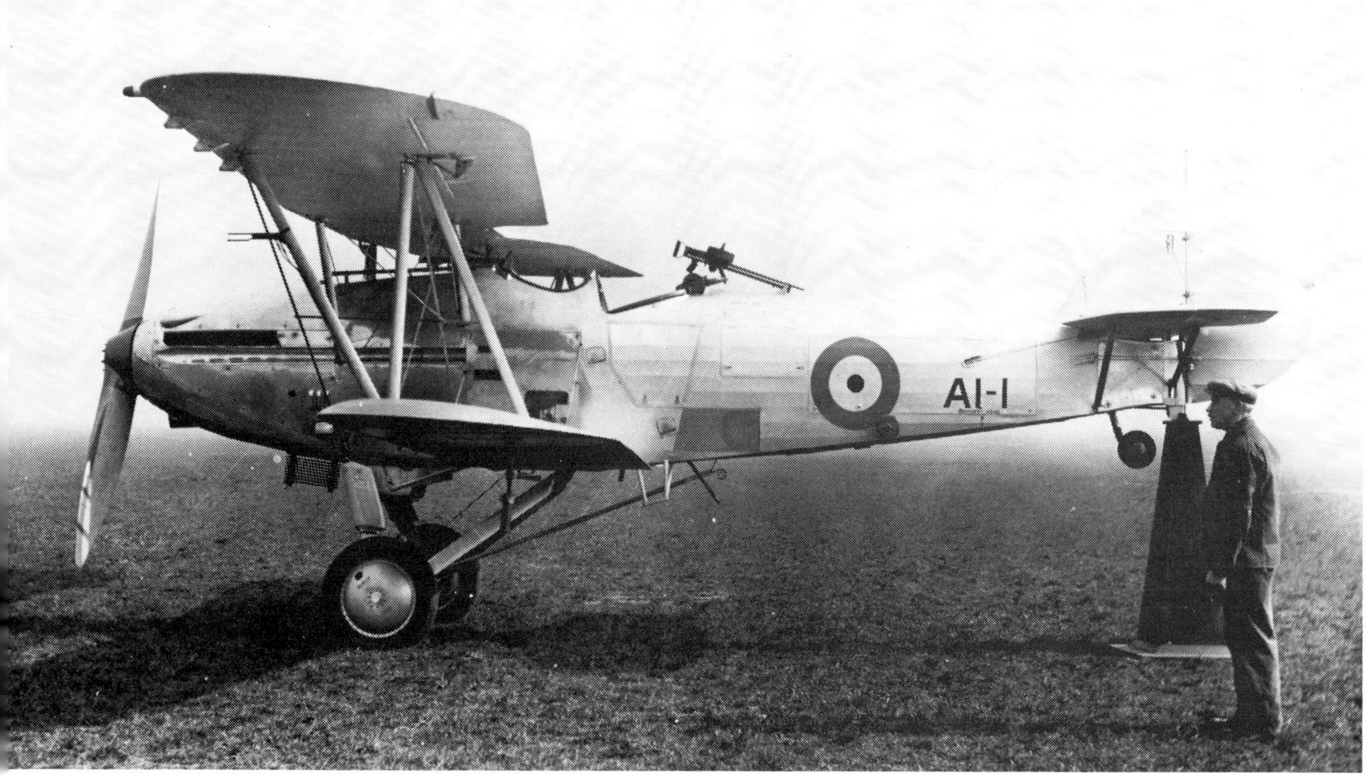
AI-I

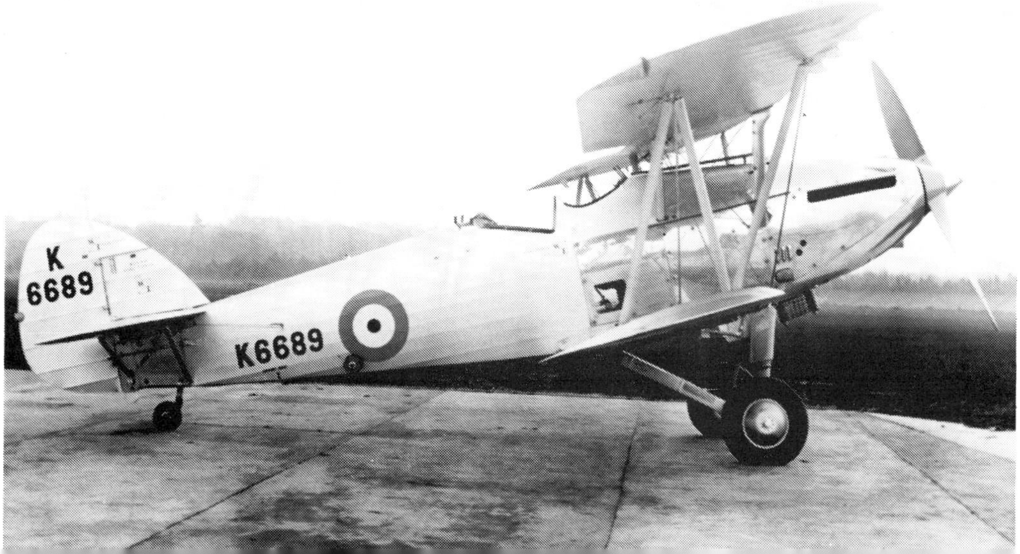
K
6689
K6689

Not only were there corporate changes but radical design changes were in the wind as well. Conceived in the fertile brain of Sydney Camm in 1933, the *Hurricane* flew in prototype form on 6 November 1935. It was just one of many hopeful aircraft submitted by various makers for approval by the Air Ministry which identified two, the *Hurricane* and the Vickers-Supermarine *Spitfire,* as the most promising types.

There were different design philosophies behind the two types. The *Hurricane* had an all metal frame with a fabric covering while the *Spitfire* was of all metal stressed skin monocoque construction. Founded on established principles the *Hurricane* could be quickly put into production and this was duly accomplished in 1936. The *Spitfire* was breaking new ground, designwise, and required a longer gestation period. *Hurricanes* were improved by fitting metal stressed skin wings in 1939.

To increase the production rate of *Hurricanes,* Hawker used shadow factories at Brooklands and Langley. Gloster, a Hawker Siddeley company, was to turn out a total of 2,750 Hurricanes from Hucclecote and a new factory at nearby Brockworth. Austin Motor Company also built 300 *Hurricanes* at Longbridge. Help was forthcoming from the Commonwealth and at Fort William, Ontario, Canadian Car & Foundry commenced production which was to reach over one thousand *Hurricanes.*

Inevitably, the *Hurricane* was to appear in many forms to keep pace with battle requirements. Up until the ending of the Battle of Britain in 1941 the emphasis had been on fighters but, after this, *Hurricanes* were increasingly used for ground attack carrying 500lb. bombs under the wings. *Sea Hurricanes* went into battle in 1940 and many 'hooked' *Hurricanes* were conversions of existing types.

When *Hurricane PZ865* named "Last of the Many" was rolled out in July 1944 it brought the total production of the breed to 14,527. By this date the *Spitfire* was the front line fighter but Hawker was not about to leave it that way any longer than necessary.

The need for an eventual *Spitfire* replacement had in fact been recognised in 1937 when the Air Ministry issued specification F.18/37 to initiate ideas. Sydney Camm proposed two designs; an 'R'-type with a Rolls Royce *Vulture* 'X'-formation engine (became *Tornado*) and an 'N'-type having a Napier *Sabre* 'H'-arrangement engine *(Typhoon).* The prototype *Tornado, P5219,* first flew at Langley on 6 October 1939 but Hawker experienced the same engine problems as Avro did with the *Vulture*-engined *Manchester.*

Only four *Tornado* prototypes were built. First flown on 24 February 1940, the *Typhoon* was better and went into production at Hawker and Gloster factories.

In the search for improved speed and rate of climb a *Typhoon Mk.II* was devised with a thinner elliptical wing and, renamed *Tempest,* was first flown with a *Sabre* engine. Problems with the *Sabre* resulted in the Bristol *Centaurus*-engined *Tempest II,* which first flew on 28 June 1943, going into full scale production in fighter and fighter bomber versions.

With this family of machines Hawker adopted all metal stressed skin construction except for the front fuselage which retained tubular framing to facilitate engine variations.

September 1944 saw the first flight of a lightened version of the *Tempest* having a shorter wingspan and known as the *Fury.* With the Bristol *Centaurus* engine installation well proven it was developed as a navalised version, *Sea Fury,* which remained in production until 1948.

Below:
HAWKER HURRICANE
Merlin engined Hurricane IIB. The Hurricane was undoubtedly Hawkers best ever aircraft in terms of sales volume with no fewer than 14,527 being built by Hawkers and licencees.

Above:
EJ743, a Tempest V series 2, with Napier Sabre engine, being piloted by Bill Humble. Almost 1,400 Tempests were built in various forms including 50 by Bristol at Weston Super Mare.

Below:
HAWKER SEA FURY
Sea Fury FB.10 with Bristol Centaurus engine. TF952 was typical of the Sea Fury fighter bombers with folding wings supplied to the Royal Navy from 1946 onward. Sea Furies were exported to the Netherlands, Germany, Iraq, Canada and Pakistan.

Jet propelled fighters were first considered by Hawker in 1943 when schemes were prepared around the Rolls Royce B.41 (Nene) turbojet.

Progenitor of a new breed of Hawker machines, the private venture P.1040 prototype (VP401) became airborne on 2 September 1947. Having adopted the Gloster *Meteor* the RAF was little interested in the P.1040 but it found favour with the Royal Navy as the *Sea Hawk.* The prototype *Sea Hawk* flew for the first time in 1951. Production followed at Kingston but after the thirty-third aircraft the work was transferred to Armstrong Whitworth at Coventry where 500 were built.

Sopwith and Hawker had used Brooklands airfield in the early years but in the expansion prior to the 1939 war flight testing had been transferred to a new airfield at Langley in Buckinghamshire. A factory had also been established there for the *Hurricane.* With the development of Heathrow close by as the prime international airport for the United Kingdom, Langley began to be restricted so a move was made in 1951 to a former wartime base at Dunsfold, Surrey.

For a Meteor replacement Hawker built the P.1067 using the then new Rolls Royce AJ.65 (Avon) turbojet and having a tail exhaust instead of the wing root outlets of the *Sea Hawk. WB188,* the prototype P.1067, flew on 20 July 1951 and was followed on 16 May 1953 by the flight of the first production *Hunter.*

Production of the *Hunter* was entrusted to Hawker works including the Blackpool shadow factory and Armstrong Whitworth in Coventry. Valuable export earnings resulted from licencing agreements with Avions Fairey in Belgium and Fokker of Holland. The Hunter was Hawker's most successful post war type with 1,972 being built. The aircraft were very popular with aircrews and are finding a new lease of life on retirement from service being sought-after by the more affluent private flyers.

In 1956 Monsieur Wibault, a Frenchman and owner of Avions Michel Wibault, approached Bristol Aero Engines Limited with a proposal for a vectored thrust aircraft power unit based upon the Bristol *Orion* turboprop engine. His idea was unworkable but was developed using a Bristol *Orpheus* lightweight turbojet having an overhung fan of *Olympus* origin blowing air out of two swivelling nozzles. Sydney Camm suggested using the Hawker patented

Left Upper:
HAWKER SEA HAWK
Coventry-built Sea Hawk F.G.A.6 with R.R.Nene-103 engine (XE456). The first 33 Sea Hawks were erected at Dunsfold but the main production run of 500 aircraft was undertaken by Armstrong Whitworth Aircraft.

Left Lower:
HAWKER HUNTER
Hunter F.6 built at Kingston (XG129) shown with Bristol patented plastic drop tanks for extra fuel. The Hunter was Hawkers greatest success in post war years and 1,972 were built in Britain, Belgium and Holland in the 1950s.

Above:
HAWKER P.1127
The second prototype P.1127 (XP836) at Dunsfold. Powered by a Bristol Siddeley Pegasus 2 vectored-thrust turbofan it first flew in 1960. *B.E.D.*

Below:
KINGSTON UPON THAMES
A 1976 view of the former Hawker factory in Richmond Road. Originally the Ham Works of Sopwith Aviation (built during WW1), the front office and rear workshop extensions were added in the 1950s when Leyland Motors' tenancy ended.

bifurcated exhaust as used on the *Sea Hawk* to give four nozzles. By cross fertilisation of ideas the Bristol *Pegasus* vectored thrust turbofan was born and first ran on the 'bench' in 1959.

A suitable airframe design, P.1127, was worked out at Kingston and the prototype P.1127 *(XP831)* flew under tether on 21 October 1960 at Dunsfold and achieved free flight on 19 November 1960. Nine pre-production P.1127 *Kestrel* aircraft were built for operational trials by the Tripartite Evaluation Squadron made up of Service crews from Britain, Germany and the United States in 1964.

The P.1127 was a private venture and adequate financing of the project by the British Government could not be obtained. Through the Mutual Weapons Development Program the U.S. Government financed the aircraft that was to evolve into the world-beating *Harrier*. As the 'jump jet' began to be recognised as an ideal battlefield weapon, political decisions had a drastic influence on the British aircraft industry. Development of the *Harrier* will be dealt with in later chapters.

FOLLAND
...And Other Hamble Aircraft

FOLLAND AIRCRAFT LTD

The history of the Hamble works of British Aerospace is essentially that of the aircraft company founded by H.P. Folland in 1937 from which it is directly descended.

Compared with the big names in British aviation, Folland Aviation produced relatively few types of aircraft of its own design and manufacture. However, Hamble has been deeply involved in aviation since the First World War. Mention must, therefore, be made of the Avro, Fairey, Armstrong Whitworth and other aeroplanes that were produced on the shores of Southampton Water and also the work of Air Service Training Limited (A.S.T.).

Taken together these individual endeavours have placed Hamble firmly in the pages of the aviation history books.

Most people know Hamble as a small picturesque yachting village just down Southampton Water from Southampton and guide books rarely mention its main claim to fame — aviation.

Before World War One, Admiralty float and seaplanes had been assembled at a Government Establishment on Hamble Point where Hamble River joined Southampton Water.

One of Hamble's boat builders, Luke & Co. built one seaplane to its own design "Admiralty Machine 105" in 1912.

Other early aviators have been recorded as making their flights from "a field at Hamble" but exactly which field and whether the same one was used each time is not known.

Fairey Aviation Company

This company was the first to set up business in Hamble using the Admiralty sheds at Hamble Spit which were rented and later purchased. The company was first registered in July 1915 at Hayes in Middlesex and the Hamble site was used at first for the assembly of seaplanes.

The first aircraft flown in 1916 were a batch of Short 827 seaplanes No's 8550-61 and these were followed by one of Fairey's own design, the *Campania*, a carrier borne seaplane which first flew on 16th February 1917. Eventually some 60 further aircraft of this type were built and assembled at Hamble.

Other small batches of various types of seaplanes were built over the next few years but land planes built at Hamble were taken to Avro's or A.S.T.'s airfield for test flying and then sent away by rail from Netley Station.

Some recorded first flights from Hamble's airfields of Fairey aircraft include: 28th November 1922 Fairey *Flycatcher, N163,* 5th November 1936 *Seafox, K4305,* and 15th October 1948 *Primer* trainer *G-ALBL.*

In 1932 a fire destroyed most of the old Admiralty sheds and the factory had to be rebuilt. During the 1930's most of the work consisted of building components for various prototype and production aircraft but floatplanes still used the slipway. In the war years production was of components for *Swordfish, Albacore* and *Fulmar* aircraft in the main.

A hangar was built on the south west corner of AST's airfield in 1941 and was used by Fairey as a flight shed for *Swordfish* aircraft after overhaul and modification. By 1950, *Firefly* and *Gannet* aircraft used it whilst the now preserved *Swordfish LS326/G-AJVH* was test flown from there in October 1955.

Aircraft production continued at Hamble until 1958 but from that time onwards the production of small boats was the main source of employment for the somewhat reduced staff. The site was sold circa 1984 to Cougar Marine after a planning application to build blocks of flats there was refused.

A.V. Roe and Co. Hamble (South) Airfield

In 1916 a site of some 300 acres was purchased by A.V. Roe and Co on the eastern bank of Southampton Water at Hamble where a factory, airfield and housing estate were planned. The factory and airfield were sited between Hamble Lane and Southampton Water whilst the housing estate was planned to be a half-mile or so away to the north. There was a slipway on Southampton water for floatplanes and the airfield had rather a steep slope towards the water from the factory, the slope being into the prevailing wind.

Above:
H.P. FOLLAND 1889-1954
Henry Phillip Folland photographed in 1950. Starting in aircraft work in 1912 at Farnborough, he moved to Nieuport in 1917 when the design team at the Royal Aircraft Factory was disbanded. He joined Gloucestershire Aircraft in 1921 as chief designer and was responsible for the Gauntlet and Gladiator fighters of the 1930s. In 1938 he was appointed a director of British Marine which was later renamed Folland Aircraft Limited.

Left:
FOLLAND GNAT
Gnat T.I two-seat trainer, XP541, with Bristol Orpheus turbojet engine on a test flight from Chilbolton, Hampshire.

The text for this chapter is based on an original manuscript by **Tony Sedgwick.**

Above:
ENSIGN
The prototype Armstrong-Whitworth Ensign, G-ADSR, on Hamble North Airfield probably on the day of its maiden flight in January 1938. Fourteen A.W. Ensigns were eventually built at Hamble.

Below:
CUTTY SARK
Saunders-Roe A.17 Cutty Sark, G-ACDR, taxying in Southampton Water with an Air Service Training launch in attendance. Picture taken circa 1933. *Rolls Royce.*

About the same time as the Avro factory was being built, an Admiralty seaplane base was being constructed just down stream. The barracks for this site were constructed on the north side of Hamble Lane (B3397) and a railway line was built from the L.S.W.R. Southampton — Portsmouth line to bring in materials etc, and a branch was built from this line into the Avro factory. This base was never completed and laid derelict until 1925 when Shell Mex and BP Ltd took the land over for a petrol and oil depot.

Production soon started in the new Avro factory, the Avro 504 being built in quantity but when hostilities ceased there was a change to the production of prototype and experimental aircraft. Bert Hinkler was for some time a test pilot at Hamble on these prototype and experimental aircraft.

The airfield and slipway were also used by other companies for test flying, an example being the Vickers *Viastra, G-AAUB,* a landplane built at Woolston and shipped down to Hamble on a pontoon which was a former Woolston-Southampton Floating Bridge No 7; this sank on 8th March 1928 after a collision with a tug and had subsequently been raised. *G-AAUB* first flew from Hamble on 1st October 1930.

During 1925 Don Juan de la Cierva brought his C6A *Autogyro,* which used an Avro 504 fuselage, to Hamble from Spain and later the same year the Air Ministry ordered a similar machine from Avro. Further orders followed and several experimental autogyros were built and test flown from the Avro site. Work continued on a general experimental category until 1932 when the airfield and factory were placed on a care and maintenance basis with just a few odd aircraft built in Manchester being assembled at Hamble, these included a batch of Avro *Cadets* for A.S.T.

Above:
EMPIRE BOAT
Imperial Airways' Short S23 Empire C-class flying boat "Cambria", G-ADUV, which flew on trans-Atlantic trials in 1937. Afterwards, it was converted for passenger carrying by Folland at Hamble as seen here. It had Bristol Pegasus engines.

In 1934, the derelict factory was re-activated by Sir W.G. Armstrong-Whitworth Aircraft which extended it prior to starting work on a new 123 ft wing span airliner, the A.W.27 *Ensign* class. The prototype, *G-ADSR*, some 18 months behind schedule, first flew from the 1926 North airfield on 24th January 1938. It had previously crossed the road from the factory to airfield under its own power. In all fourteen of these aircraft were built, the last one *G-AFZV* 'Enterprise' leaving on 28th October 1941.

The second type built was a prototype wood and steel bomber with tricycle undercarriage, the A.W.41, later known as the *Albemarle*. It first flew from A.S.T.'s airfield on 20th March 1940 and was numbered *P1360. P1361,* the second *Albemarle* was also built and flown from Hamble but subsequent production was by various sub-contractors including A.W. Hawksley at Stroud, Glos., and the type was not seen at Hamble again.

Parts for *Whitley* bombers and other Hawker Siddeley aircraft kept a small staff busy in the early war years and a number of Handley Page *Hampden* bombers were converted to torpedo bombers during 1940. From August 1940 the factory became part of Air Service Training Ltd.

Air Service Training Ltd, Hamble (North) Airfield
During 1926 Avro purchased farmland to the north of its houses in Verdon Avenue up to the Southampton-Portsmouth railway, this becoming their 'North' airfield, the original site becoming rather too small for general use and was relegated for use for smaller types and autogyros.

In August 1926 the Hampshire Aero Club was formed using a small group of buildings on the north east of the 'barrack' complex for hangarage and clubrooms and the new airfield for flying. Their fleet originally consisted of *Cirrus Moths* but later a *Gypsy Moth G-ABBD* and Avro *Avian-III G-EBVI* were added to the fleet before the club moved to Eastleigh in November 1932.

In 1931 the Armstrong-Whitworth Training and Reserve School moved from Whitley to new hangars built on the airfield and became Air Service Training Limited, formed in February to commence flying in April and officially opened in June. The aircraft types available at the opening included Avro 504, *A.W. Siskin, A.W. Atlas* trainers, Avro *Tutors* and *Avians.*

A.S.T. continued to expand during the 1930s gathering in students from all over the world, for civil training while the Reserve School (renamed no.3 Elementary and Reserve Flying Training School in 1933) became increasingly involved in military flying training.

At the outbreak of war, training activity increased at separate schools covering military flying, wireless instruction and navigation training. Hamble was a very busy airfield at this time but this was short-lived. Wireless training ceased and the other schools beat a hasty retreat to Watchfield on 20 July 1940 after the fall of France, and German aircraft had been seen in the area day and night.

Returning to the constructional side, when A.S.T. took over the management of the Armstrong Whitworth site on the south of Hamble Lane in 1940, some work was still outstanding on the *Ensigns* and there were some *Hampdens* to convert.

During 1941 four new hangars were built, the one

in the south west corner being used by Folland Aircraft and Fairey Aviation as a flight shed. The northerly one became No.15 Ferry Pilots Pool of the Air Transport Auxiliary which used *Ansons, Argus* and *Warferry* aircraft for pilots collecting and delivering aircraft in southern England. The one in the southern corner of the airfield was used for *Spitfire* repair work whilst one in Badnam Copse became a trial installation depot for American *Mohawk, Tomahawk, Airacobra* and *Mustang* aircraft.

After the destruction of the Supermarine Works at Woolston by an air raid in September 1940, the aircraft division of A.S.T. expanded rapidly and undertook repairs and salvage work on Spitfires.

This work continued throughout the war years with some 3,500 being dealt with. Conversion work was also undertaken and some 'hooked' *Spitfires* were modified at Hamble; later a batch of *Seafires* was delivered and the first conversion, *MB328*, flew in May 1942.

During 1941-42, the surviving *Ensigns* returned to Hamble to have more powerful engines installed, thus becoming *Ensign II's*. Early in 1943 some Boeing B17 *Flying Fortress* aircraft arrived for radar installation and other mod's whilst several North American *Mitchells* also arrived for trials installation work.

Towards the end of the war, a batch of Avro *York* transports were converted for V.I.P. use and others followed for civilian use by BOAC; the first BOAC machine flying in its civilian state on 31 August 1945. Military work gradually ran down and various odd jobs were found for the depleted work force such as conversion of *Mosquito* aircraft for the French Air Force and overhaul work on *Anson XIX's* for Railway Air Services. The 1939 built hangar was handed back to the new post war Training Division but the 1931 hangar was retained until about 1958. A large number of *Spitfires* and the remaining *Ensigns* were cut up in 1946-47.

The experimental department undertook several test bed jobs on *Lancasters* and a *Meteor* and *Hunter* as well. Other work included the building of towed targets and stand off bomb test vehicles as well as the installation of a Rolls Royce *Derwent* jet engine in *Lincoln SX971* during 1949.

Above:
SPITFIRE
EP751 was one of three Spitfire mk.V's converted with Supermarine-designed floats by Folland Aircraft. It first flew in this condition in April 1943 and was sent to Egypt in October 1943 where it flew from the Great Bitter Lake.

The Berlin airlift brought more *Yorks* to A.S.T., this time for overhaul of RAF aircraft. The Company also leased the old Avro slipway to Aquila Airways which was using *Hythe* flying boats on the airlift and also loaned staff to do repair work until Aquila formed its own workforce.

1949 saw the last aircraft in the main hangar of the old Avro site as the whole floor was laid out for the production of *Meteor* components, whilst two new hangars were built on the old airfield for Gloster *Javelin* production.

As *Javelin* production tailed off in 1959, aircraft work at A.S.T. gradually ceased until the factory became part of Petters. The Aircraft Division of A.S.T. closed down early in 1960.

The Training Division or Flying School restarted operations in 1946 with *Tiger Moths, Ansons, Whitney Straight, Proctor* and *Oxford* aircraft. Military training recommenced in March 1947 when two *Tiger Moths* of Southampton University Air Squadron arrived from Eastleigh. Training continued up to a peak in 1953 after which it went into decline.

During 1959, a BOAC/BEA training scheme was established at Hamble as the College of Air Training. It opened in 1960 and the A.S.T. name moved to Scone airfield in Perth, under Airwork Limited.

Simmonds Aircraft Limited
One of the smaller firms which used the new Hamble airfield in 1928 was Simmonds Aircraft Limited which assembled and tested its *Spartan* aircraft there. The prototype, *G-EBYU*, first flew in July 1928 and some 49 *Spartans*, built at Weston near Woolston, were flown at Hamble. The other types test flown at Hamble were the *Spartan Arrow* and the *Spartan Three Seater*. The name changed to

Spartan Aircraft Limited in 1930 and moved to Cowes, Isle of Wight in February 1931.

Folland Aircraft Ltd

During 1935 yet another aircraft factory was built in Hamble, next door to the well known military hospital at Netley. This was opened by British Marine Aircraft in 1936 to build Sikorsky Flying Boats under licence. They did not get very far with production, only one aircraft being partially completed, *G-AEGZ*, known as BMI, but was in fact a Sikorsky S.424, and the future did not look too bright.

However, Mr H. Folland arrived on the scene from Gloster and the factory became Folland Aircraft Ltd, and was soon on subcontract work for various firms. It made use of the slipway by obtaining a contract for servicing "C" class flying boats for Imperial Airways.

The first complete aircraft to be built at Follands was a batch of twelve single engined test beds to Air Ministry specification 43/37 and serialled *P1774-85*. Various engines were fitted such as *Sabres, Hercules* and *Centaurus,* but the arrival on the scene of a number of outdated *Battles* stopped further orders. The 43/37's were taken by road to Staverton near Cheltenham for completion and test flying. War time activities included production of *Spitfire* components and three *Spitfire* float planes were built complete, the first *W3760,* flying from Southampton Water in October 1942.

Various post war projects did not get beyond the mock up stage but component manufacture continued for other companies and there were excursions into commercial non-aircraft work.

In 1950 W.E.W. Petter joined the firm from English Electric and design work on a new lightweight fighter was started, the prototype known as the *Midge* first flew from Boscombe Down on 11 August 1954. Seven slightly larger aircraft, known as the *Gnat (XK724/39-41/67/8 and XN326)* were built for evaluation by the M.o.S. No British orders for the fighter were placed but 25 were supplied to the Indian Government and a licence was granted to build further aircraft. Thirteen others were supplied to Finland and two others to Yugoslavia.

In 1957 a tandem seat trainer version was proposed, to be known as the *Gnat Trainer* and during 1958-9 fourteen were built, XM691-8, XM704-9, the first flying from Chilbolton Airfield on 31 August 1959.

After Follands had joined the Hawker Siddeley Group, the firm was awarded a production order. All *Gnat Trainers,* including those for the 'Red Arrows' aerobatic team, were built complete at Hamble and test flown from Chilbolton and later Dunsfold airfields.

Right:
FOLLAND F43/37
Sabre-engined F43/37 test bed aircraft, P1774, pictured at Folland's Experimental Department at Staverton, near Cheltenham in August 1941.

Below:
FOLLAND F43/37
P1775 with Bristol Centaurus engine at Staverton. It was one of twelve airframes built for the Air Ministry for use as engine test beds and was tested with a Bristol Hercules VIII engine in March 1942.

Above:
FOLLAND MIDGE
Folland Fo139 Midge with Armstrong Siddeley Viper turbojet as first flown on 11 August 1954.

Right:
EJECTOR SEAT
Folland fully automatic ejector seat on show at Farnborough in 1960. The SAAB-inspired Folland Mk.4GT seat was fitted in Gnat trainers and achieved worldwide sales of over 700 units.

Below:
GNAT T.I
Production of Gnat Trainers at Folland's Hamble factory in the early 1960s.

Above:
INDIAN GNAT
Folland Fo141 Gnat F.I sold to the Indian Air Force (serial IE1062) in 1958. Gnat airframes and the Bristol Orpheus engine were both later built under licence by Hindustan Aircraft in Bangalore.

Above:
HAMBLE
A recent view of Hamble looking east with the BAe factory (ex Folland) in the foreground and the slipway leading from the original hangar down to Southampton Water. Hamble river is in the far distance behind AST's 'Hamble Airfield'. Armstrong Whitworth's old factory, ex Avro, is at top right and Fairey was nearer to Hamble Point off right of the picture.

English Electric Aviation

To trace the progress of English Electric Aviation at Preston, Samlesbury and Warton in Lancashire, now part of the Military Aircraft Division of British Aerospace, the pages of the calendar must be turned back to 1863.

In that year the North of England Railway Carriage & Iron Company established a works, later to be known as the East Works, at Preston. At that time the factory was situated on the bank of the River Ribble, the course of which was diverted in 1892 for the construction of Preston Docks.

A private partnership, Dick, Kerr & Company, was formed in 1875 to construct tramway and railway equipment in Kilmarnock, Scotland and was incorporated as a public limited company on 31 May 1890.

Dick, Kerr & Company Limited acquired the disused factory on the east side of Strand Road from the railway company in 1897 for the manufacture of trams which were to be exported worldwide. To manage the factory a new subsidiary company, The Electric Railway & Tramway Carriage Works Limited, was formed in 1898.

Across the road another factory, later known as the West Works, was built in 1899 by The English Electric Manufacturing Company Limited for making electrical equipment for tramways and railways.

In 1903, Dick, Kerr & Company Limited acquired control of The English Electric Manufacturing Co. Ltd., and two years later a new company called United Electric Car Company (an amalgamation of the 'Carriage Works' with Geo. F. Milnes & Co., and The British Electric Car Company Limited) was registered to run the East Works.

Production of trams continued at Strand Road until 1914 when war work was allocated including the manufacture of guns and shells. To accommodate the extra load the work force swelled to 8,000 people.

Some aircraft work was undertaken by Dick, Kerr's in 1917 and included the *Felixstowe* flying boat. A batch of fifty *Felixstowe F.3* 'boats (serials *N4230* to *N4279*), ordered at Lytham but to be assembled at South Shields, were followed by another eighteen of the same type (*N4100* to *N4117*) and a *Felixstowe F.5 (N4128)*. Later examples were assembled in a hangar at Lytham on the nearby Fylde coast and used the river Ribble for taxying and flight testing. One Fairey *Atlanta (N119)* was also built at Lytham.

Mention should be made here of the contribution of two companies which come into the main story later. The Phoenix Dynamo Manufacturing Company of Bradford, Yorkshire also built aircraft under licence starting with twelve Short 184's (serials *B368* to *B379*) delivered in 1916. One hundred and fifty Sopwith 7.F.1. *Snipes* were erected at Bradford which then received a string of orders for the Royal Naval Air Service. A single Fairey *Titania* was erected and was followed by four Armstrong Whitworth F.K.10's. Two further batches of Short 184's preceded seventy-six *Felixstowes* (*N4184* to *N4229* & *N4400* to *N4429*) and thirty Maurice Farman S.7. *Longhorns* (*N5330-N5349* & *N5750 - N5759*).

The Coventry Ordnance Works (C.O.W.) also built aircraft under Government contract and included 250 of the Farnborough designed R.E.8 (*C5026- C5125* & *D6701 - D6850*), known in the Services as "Harry Tates".

At the end of the First War the future prospect for Dick, Kerr & Company were not good and there was also the risk that the Government might impose heavy taxation in the form of an Excess Profits Duty. The English Electric Company Limited was formed in December 1918 to amalgamate Dick, Kerr & Co., and its subsidiaries; Phoenix Dynamo, Coventry Ordnance

Opposite:
LIGHTNINGS
English Electric Lightning supersonic fighters of the RAF.

Below:
PHOENIX FIVE
N.86 (works No.250) seen during running of its Rolls Royce Eagle engines in 1918. The similar Phoenix Cork (serial N.87; works No.251) was also built at Bradford, Yorks.

Works, Williams & Robinson of Rugby and Siemens Dynamo Works at Stafford. English Electric specialised in heavy electrical equipment including traction motors, generators and switchgear. On the mechanical side, diesel engines and turbines were built together with trams and railway locomotives. Strand Road Works became 'English Electric - Dick, Kerr Works - Preston' in the reorganisation.

Aircraft were also included in the new company's list of wares and to realise this W.O. Manning, Chief

Below:
E.E. WREN
J6973, the first of three examples of the English Electric Wren ultra light powered glider using a 398c.c. A.B.C. two cylinder motorcycle engine. Wrens competed in the 1923 and 1926 Light Aeroplane Competitions at Lympne and one is now preserved.

Lower:
ENGLISH ELECTRIC AYR
N148, the sole M3 Ayr experimental aircraft at Lytham in 1924 with 450hp Napier Lion IIb engine.

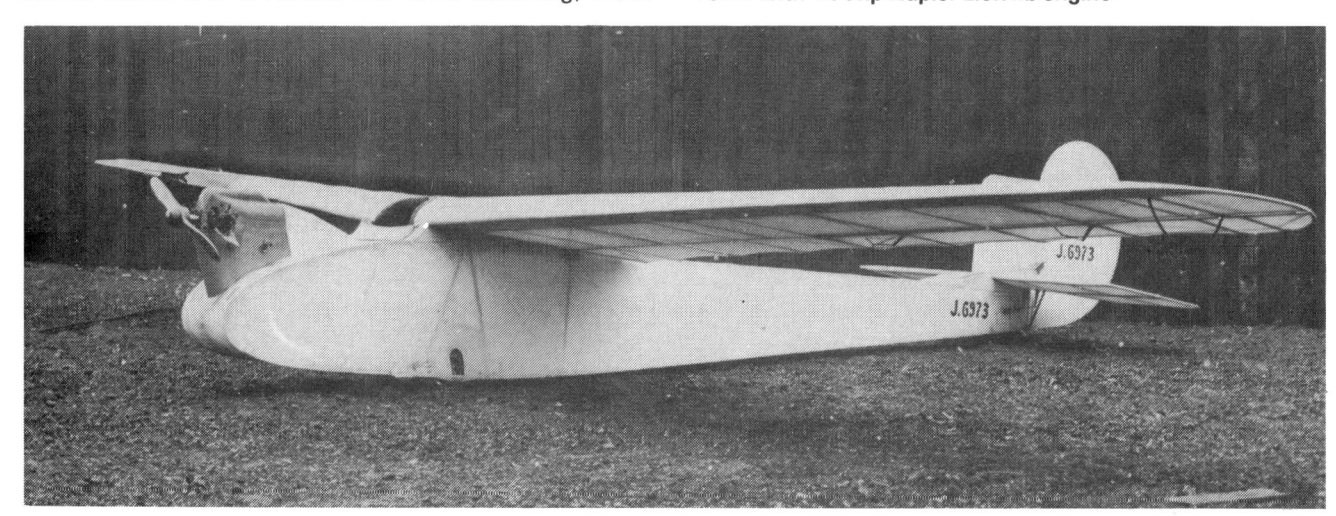

Designer of Phoenix Dynamo, set up a design office with five subordinates at the English Electric Head Office at Queen's House, Kingsway in London. Manning's team was prolific in ideas but of the thirty or so schemes that were worked out between 1919 and 1926 only three types were actually built and, of these, only two were ever flown.

Built for the Air Ministry in 1921, the prototype of the English Electric Wren single-seat, ultra-light 'powered glider' weighed a mere 420lb (190kg) and had a maximum speed of 50 mph. For the 'Daily Mail' Light Aeroplane Competition held at Lympne in 1923, two more Wrens were built at Preston and were flown by Flt. Lt. Langton and Sqdn. Ldr. Wright. The third Wren was registered G-EBNV in 1926 for entry in the 1926 Lympne trials and was subsequently stored. In 1957, one good machine was rebuilt by English Electric using parts of the two 1923 built Wrens and this was restored again in 1981 for preservation by the Shuttleworth Trust.

The English Electric M.3 *Ayr* fleet gunnery spotter and reconnaissance biplane was unusual in that the lower pair of wings was made watertight and buoyant to act as stabilisers when taking off and landing on water. Two Ayrs (N148 and N149) were commenced in 1924 but there is no record that either one was flown.

English Electric had greater success with the *Kingston* anti-submarine and coastal patrol flying boat. Developed from the Phoenix *Cork* of 1918, the

Above:
KINGSTON
English Electric P5 Kingston flying boat (N9710) at Lytham, Lancashire. The engine nacelles were extended aftwards to provide gunners' stations.

first machine designated type P.6. *Kingston (N168)* had a double skin wooden monocoque hull but the second machine had a metal hull. Five P.5 *Kingstons* were built serialled *N9709* to *N9713* and all had Napier *Lion* engines. Like the *Ayr* they were built at Preston and assembled at Lytham.

Throughout its history English Electric Company was a highly profit oriented concern and if a product became only marginally profitable it was soon dropped. After the brief flirtation with aircraft building in the decade from 1916 the aircraft department was closed and E.E. concentrated on its heavy engineering and electrical interests.

As the incurable aircraft constructors struggled to survive on meagre orders yet with increasing costs, English Electric found its mark in supplying locomotives to the 'Big Four' railway companies. The diesel-electric six-wheeled shunter was adapted to suit the needs of the individual users and the acceptance of the English Electric product as the standard for the industry ensured massive orders until well after the Second War.

On the eve of what was to become the first war to be dominated by aircraft in the order of battle, Preston was called upon by the Government to turn its efficient manufacturing philosophy over to aircraft production once again.

1938 therefore saw Preston engaged in the large scale production of the Handley-Page *Hampden* and later the *Halifax*, both bombers. Preston being situated in a built-up area, it was necessary to set up a final assembly and flight test facility. This was duly established at Samlesbury between Preston and the town of Blackburn to the east on the site of the proposed Preston/Blackburn airport; (not to be confused with the company of that name founded by Robert Blackburn).

The first seventy-five *Hampden I's* were ordered in 1938 and the first of these, *P2062*, made its maiden flight from Samlesbury on 22 February 1940. These were followed by the first batch of two hundred *Halifax II* bombers *(V9976-V9994)* and *W1002-W1276* with 'blackout blocks') and *V9976* became airborne at Samlesbury on 25 August 1941. Thereafter, large batches of each type followed until 900 *Hampdens* and 2145 *Halifaxes* had been built.

While these standard bombers were leaving Samlesbury at a rate of up to eighty per month preparations were being made under great secrecy to build a new breed of fighter aircraft, the jet propelled *Vampire*. Built to Air Ministry specification E6/41 the first of two prototypes *LZ548/G* was built by de Havilland and flew at Hatfield on 20 September 1943.

In 1944 English Electric received orders for 120, later increased to 300, *Vampire F.Mk.I.* and the first of these, *TG274/G*, flew at Samlesbury on 20 April 1945. Preston and Samlesbury were eventually to be responsible for 1366 *Vampires* and the experience gained in 'jet' aircraft was to have irreversible consequences much to the chagrin of the established aircraft firms.

Between 1940 and early 1942 when it was formally handed over, a new air base had been built by the Government at Warton near the old Dick, Kerr base at Lytham. Staffed by 20,000 American servicemen, mostly technical trades, Base Air Depot No.2 was part of the 'Air Bridge' bringing aircraft from the United States to Britain for action against Germany.

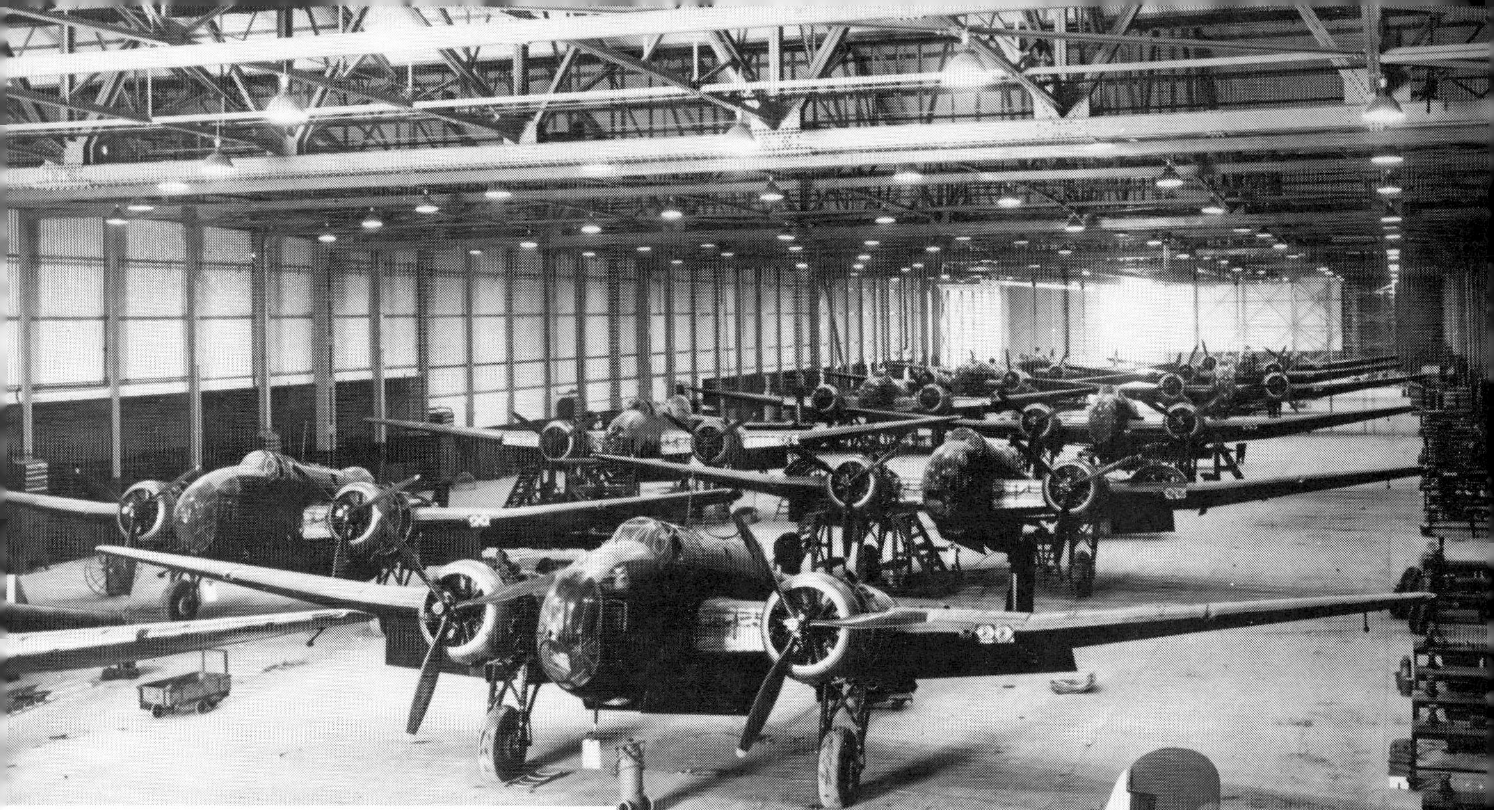

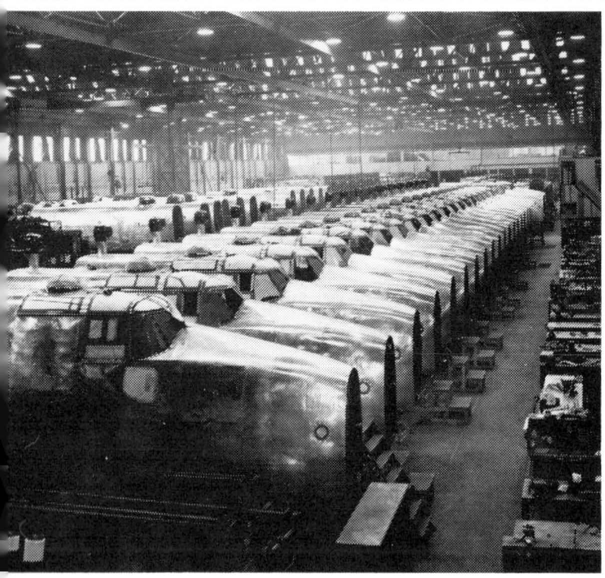

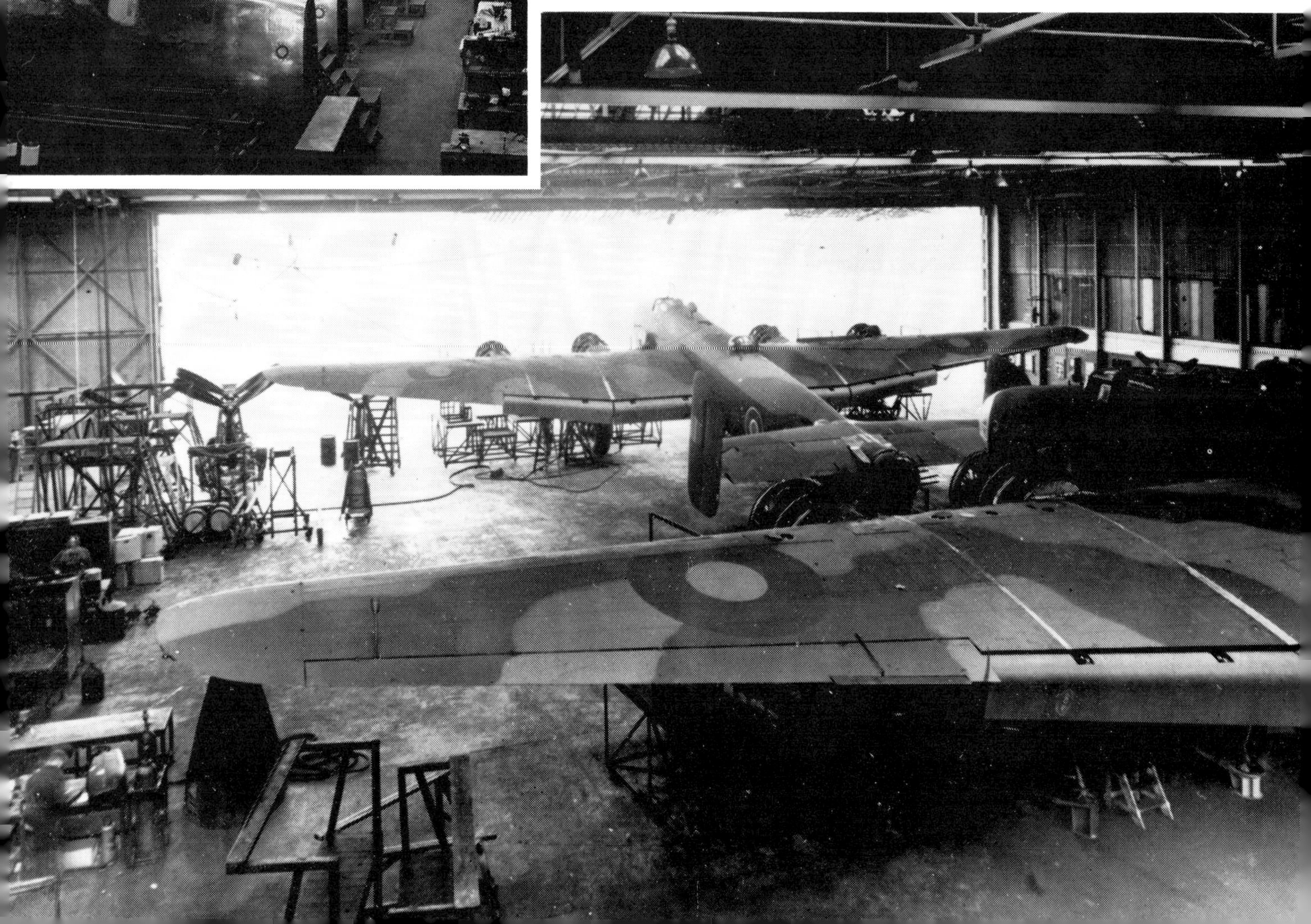

Left Upper:
HAMPDEN
Handley Page Hampdens with Bristol Pegasus engines in production at English Electric's Samlesbury works during the 1939-45 war.

Left Centre:
HALIFAX
Halifax fuselage sections for about 25 aircraft lined up during the War.

Left Lower:
HALIFAX
Wartime production of Handley Page Halifax bombers at English Electric's Samlesbury shadow factory.

Above:
VAMPIRE F.I.
English Electric moved into the jet age when it took on the production of the de Havilland Vampire at Ministry request. TG/292 is visible in the foreground in this view of Vampire assembly at Samlesbury in the mid-1940s.

Below:
CANBERRA
This Royal Navy Canberra T.22 Radar Target was one of seven conversions of retired RAF Canberras.

During the American tenure, which did not end until Japan was defeated later in 1945, the types re-assembled and prepared for war at Warton included *Flying Fortresses, Liberators, Marauders, Havocs, Thunderbolts, Airacobras, Lightnings, Tomahawks* and *Mustangs*. Several hundred aircraft were closely parked at Warton at any one time awaiting attention; smaller fighters were even 'garaged' under the wings of the large bombers.

After the War, the American forces were demobbed and Warton was the home of several small RAF units.

Seeking to remain in aircraft work this time round the English Electric board of directors identified the jet aircraft as the type for the future. To realise this ambition, design capability was essential; something that English Electric lacked. W.E.W. 'Teddy' Petter was appointed Chief Designer and with a small team set up a design office on the first floor of a garage in Preston. They started work on the A.1 later to become the *Canberra*, Britain's first jet bomber. In 1947, design activity removed to Warton to continue the work in more secure surroundings and an Aircraft Division was formally established in that year.

One year later a force of about one hundred people arrived at Warton to make preparations for the final assembly and testing of the new *Canberra* bomber.

A low speed wind tunnel was set up in 1948 and supplemented by a transonic tunnel built in 1951 to explore flight conditions of the proposed P.1 at mach one.

The first English Electric flight testing at Warton was a high altitude research programme with a

Right:
Strand Road bisects the Preston factories with BAe (Dick, Kerr's) on the left of the picture.

Below:
Warton airfield on the Ribble estuary.

Right Lower:
Samlesbury airfield and factory were built for World War 2 production of bombers for the RAF.

Bottom:
MARTIN RB57
Canberras were built under licence in the USA as the B57. An RB57 of the US Air Force is shown taking off from Martin Aircraft's Baltimore plant in 1954. *Martin Aircraft.*

Above:
E.E. LIGHTNING
The second prototype English Electric P.1B fighter (XA853) built at Warton and seen armed with two Firestreak (nee Blue Jay) missiles built by de Havilland Propellers at Stevenage. The P.1B was christened Lightning in 1958 and the type entered RAF service in 1960.

Below:
THUNDERBIRD
The English Electric Thunderbird anti aircraft missile was built at English Electric Stevenage (now BAe Stevenage 'A') for the British Army. Thunderbird I equipped nos. 36 and 37 Anti Aircraft Regiments in 1960 and was replaced by Thunderbird II in 1965.

Gloster *Meteor 4* between June 1947 and spring 1948.

On Friday 13 May 1949 the prototype *Canberra, VN799,* flew for the first time from Warton and was joined by the second prototype in the following year. By 1950 most of the technical staff had moved to Warton from Preston.

August 1954 saw the first flight of the prototype P.1A., serial *WG760*, at Boscombe Down and on 4 April 1957 the prototype P.1B *Lightning, XA847*, flew at Warton.

Site development included a runway extension in 1956, a new control tower in 1957 and major new research and development facilities including the construction of twin high speed wind tunnels between 1956 and 1959.

The aircraft and guided weapons activities of English Electric were separated from the parent company on 20 April 1959 by the formation of English Electric Aviation Limited having an Aircraft Division at Warton and a Guided Weapons Division at Stevenage.

The influence of Warton continued to increase rapidly after 1960 but by this time English Electric had, by Government decree, become a founder member of British Aircraft Corporation; but that is another story.

... Out Of The Shadows
"CHESTER"

The 'shadow' factory at Broughton, sometimes called Hawarden has never had a corporate autonomy of its own and should, rightfully, be described in the chapters on Vickers and de Havilland. However, it has gained prominence through its high volume production record and now, as part of British Aerospace, has influence about equal to the pioneer firms in the industry.

In 1934 the British Government gave up its misguided ideas that disarmament was the road to peace and faced the reality that Hitler was building up the German arsenal ready for another war.

To meet the increased demand for aircraft extra factories were required to be built. In considering new plant it was decided that production should be dispersed as much as possible to prevent a stoppage in the event of a successful attack on one factory. As a further safeguard the new factories were destined to be built as far as possible out of range of enemy bombers. This programme became known as the "shadow factory" scheme and several new factories were built in the north west of England.

One such shadow factory was built near the village of Broughton in Flintshire just 3½ miles to the south west of the ancient city of Chester across the county boundary. With a floor area of one million square feet the Broughton factory with its adjoining Hawarden airfield was built for the Government and was ready for occupation in 1937.

Vickers was preparing to put its new geodetic bomber, the *Wellington*, into large scale production. However, Weybridge alone could not cope with the massive orders that had been placed. Sir Robert McClean, Chairman of Vickers-Armstrongs Aircraft had approached the Secretary of State in 1936 with plans for an additional new factory.

Broughton factory was therefore leased to Vickers and *Wellington I*, serial *L7770*, was the first aircraft to leave Hawarden in September 1939. Monthly output reached 130 *Wellingtons* at the peak and by 1945 a total of 5,540 had been built at this location; almost half the combined total of 11,460 *Wellingtons* from all sources.

As the strategy of the RAF changed to heavy bombing raids on German industrial targets Vickers Armstrong (Chester) was given an order in April 1943 for five hundred Avro *Lancaster B.Mk.I.* bombers. From June 1944 to September 1945 when the war ended 235 *Lancasters* had been delivered. These included fifteen aircraft transferred unassembled from the Metropolitan-Vickers factory at Trafford Park, Manchester. They were erected between 15 June and 13 August 1945 and subsequently flown to Armstrong Whitworth at Coventry late in 1946 for conversion to *Lancaster B.Mk.I.(F.E.)*.

The final World War Two order to be erected at Broughton was for eleven Avro *Lincolns* after which, with M.o.S. orders completed, the factory began to run down.

Before the war ended the Government turned its attention to the problem of housing the people made homeless by the Blitz. As there was most probably going to be a surplus of aircraft skills and materials the M.o.S. organised a competition among the aircraft firms for an aluminium prefabricated house.

The contest was won by Bristol Aeroplane Company and Broughton joined forces with the Bristol-managed shadow factory at Weston-Super-Mare, Vickers at Blackpool, Blackburn at Dumbarton and others in building the A.I.R.O.H. 'prefabs'. By 1948 Broughton had built 28,000 units and Weston 70,000. The latter site then turned to producing schools and other buildings but 'Chester' started to produce aircraft once again.

At this juncture the management of the site was taken over by de Havilland and as the last houses were being constructed at one end of the assembly shop aircraft jigs were being installed at the other. After the formal 'handing over' ceremony on 1 July 1948 to de Havilland (Broughton) production of the wooden *Mosquito* and its wood and metal counterpart, the *Hornet*, got under way. The first "Mozzie" of a batch of thirty-three *Mosquito NF.38's* for Jugoslavia was flown out in October and the first 'Chester' *Hornet* flew in March 1949. These two types were joined by an export order for *Vampire FB.5's* for Sweden and Switzerland.

The de Havilland Company of Canada had designed a rugged single engined monoplane and when this was adopted as a trainer by the RAF, production in the U.K. of the DHC.1., *Chipmunk* was handed to Chester. The *Chipmunk* line was introduced in 1950.

The early years of the 'fifties saw much diversity of de Havilland designs built at Chester. In 1951, *Dove* production was transferred from Hatfield to Broughton and this twin engined transport was to share the floor with *Vampire* night fighters. In the following year they were joined by the *Venom* (a *Vampire* derivative) and the *Vampire Trainer*.

Besides the *Dove*, Broughton also built the production batches of the larger four engine D.H.114 *Heron*. Of the 140 *Herons* built there mention should be made of the *Heron C. (VVIP)4's* built for the Queen's Flight. The factory has been responsible for making and overhauling aircraft for the Queen's Flight ever since.

By 1952, 'Chester' had demonstrated its ability to handle a variety of aircraft of all sizes from the diminutive *Chipmunk* up to the *Lincoln*. Work on the *Comet 2* airliner ordered by British Overseas Airways Corporation commenced at Broughton in that year but after the loss of three *Comets* all production was suspended in 1954 before the first Chester *Comet* had flown. After it was decided that further *Comets* would have a strengthened fuselage the Mk.2's were reworked as transports for the RAF. Chester's *Comet 2's* were completed between 1955 and 1957. Production then switched to the *Comet 4* and the first of these, *G-APDE*, was delivered to B.O.A.C. in 1959.

In January 1960 de Havilland (Chester) became part of the Hawker Siddeley Group but for the next three years the change over was not noticeable in Chester products. In 1961 and 1962 another de Havilland Canada aircraft was produced and including a repeat order in 1968 a total of forty-six DHC.2 *Beavers* was built. Production of the D.H.110 *Sea Vixen* commenced at Chester in 1962 and thirty had been built by 1968.

Opposite:
WELLINGTON III
Thousands of 'Wimpy' bombers were assembled at Broughton, near Chester, or built complete at Squires Gate, Blackpool; both 'shadow factories' being managed by Vickers at Weybridge. Wellington IIIs, shown here, had Bristol Hercules engines.

Above:
WELLINGTON
Broughton-built Wellington I (R1333) receiving close scrutiny by the workforce on 7 November 1940.

Below:
LANCASTER
Assembly of Lancasters on multiple flow lines at Broughton. Parts and sub assemblies were supplied to the line from Manchester and sub contractors.

Right Upper:
SEA HORNET
Line-up at Broughton on 28 November 1949 with VW979, a Sea Hornet NF Mk22, in the centre and a Sea Hornet F Mk20 nearest the camera.

Right:
VAMPIRE
Lebanese Vampire L157 at Broughton.

Right Lower:
D.H.115
Vampire Trainer serial 28431, seen here taking-off from Broughton on 25 August 1953, was one of forty-five T.MK55's sold to Sweden where they were known as the type J28C.

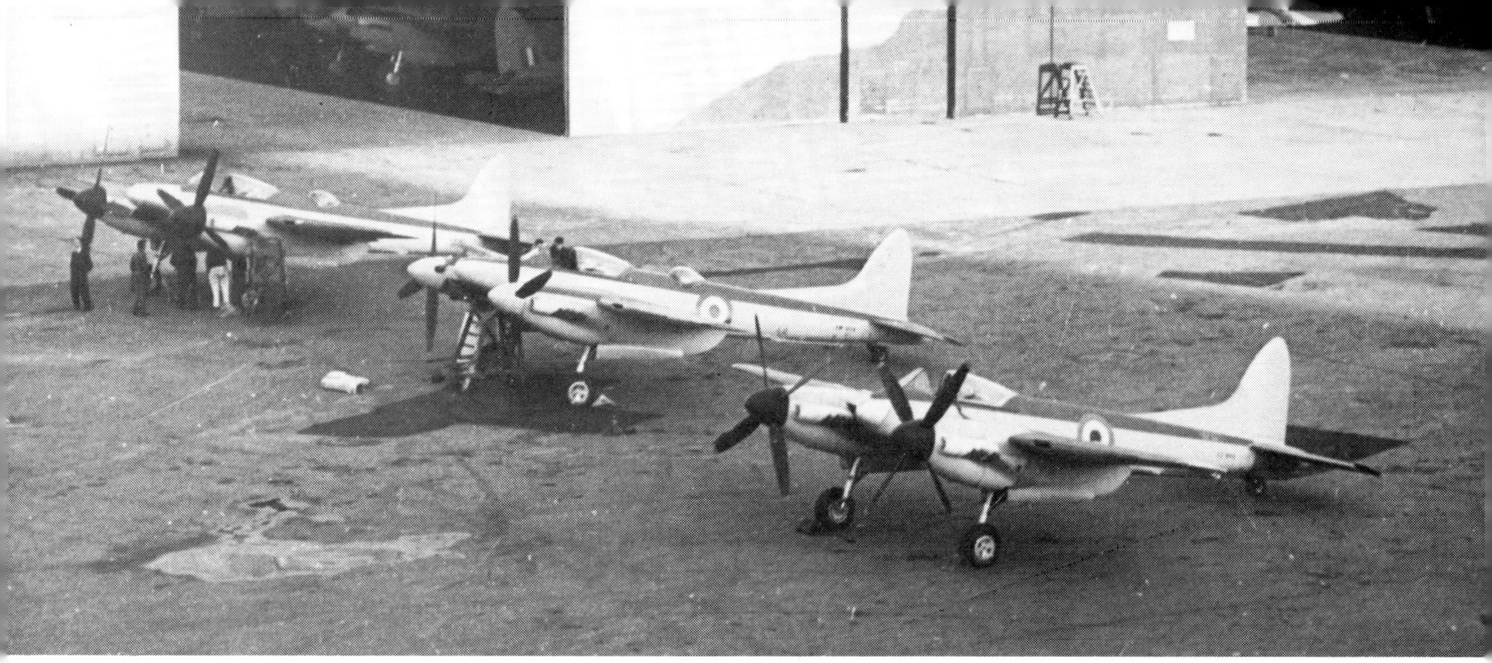

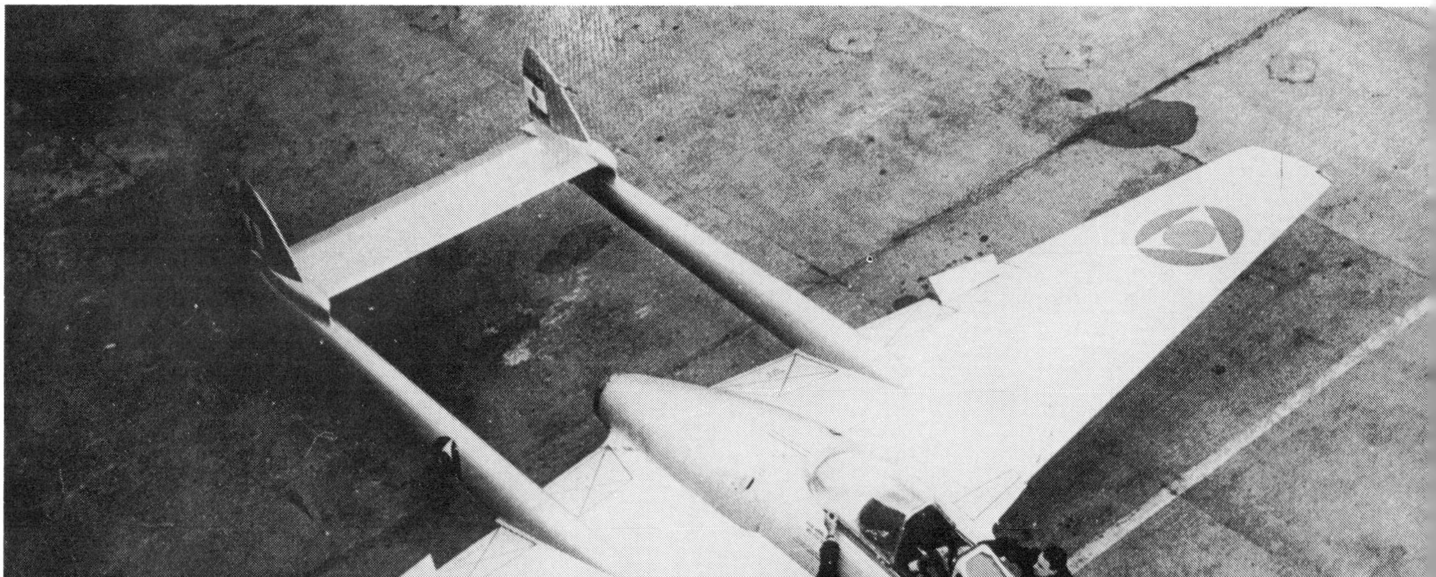
L157

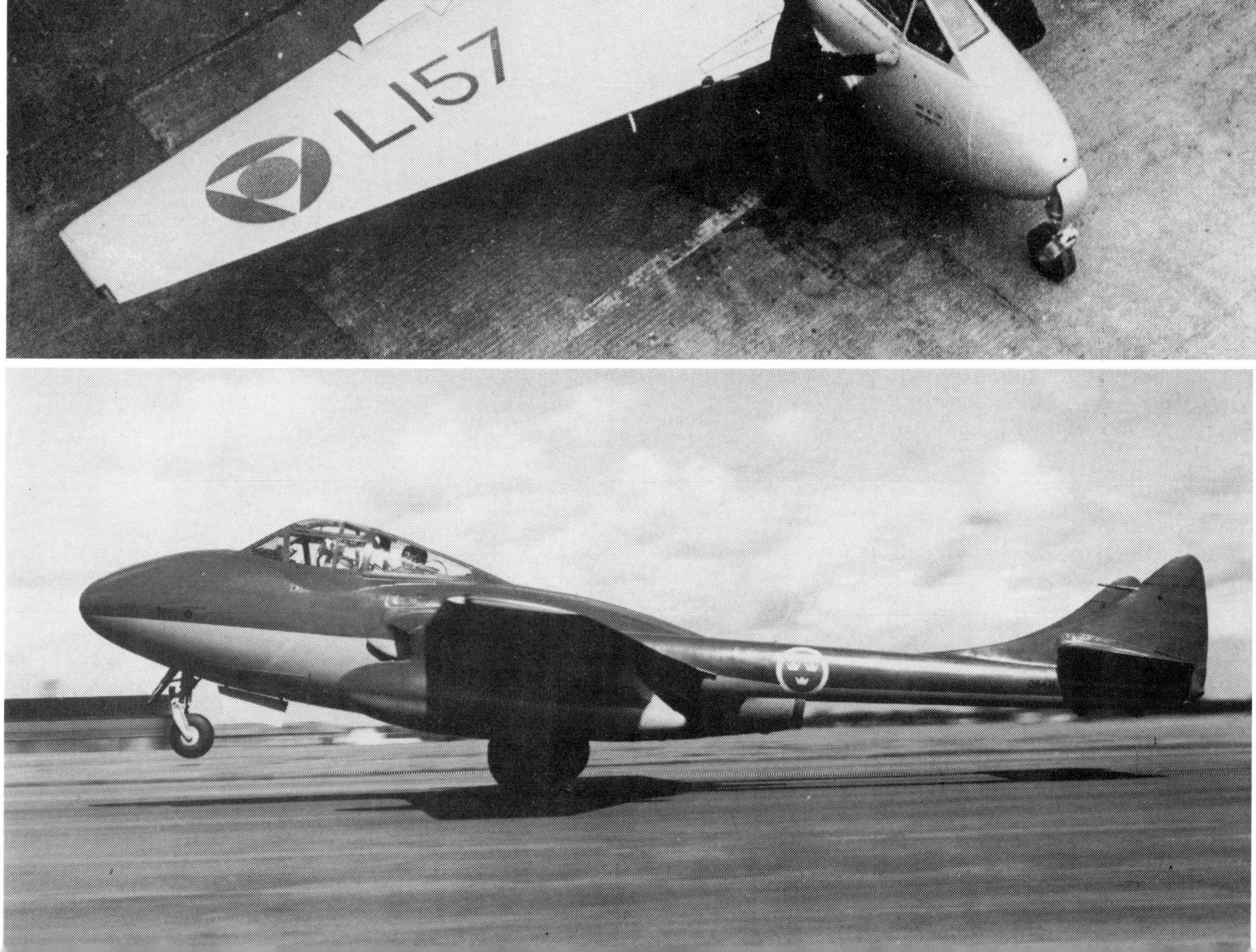

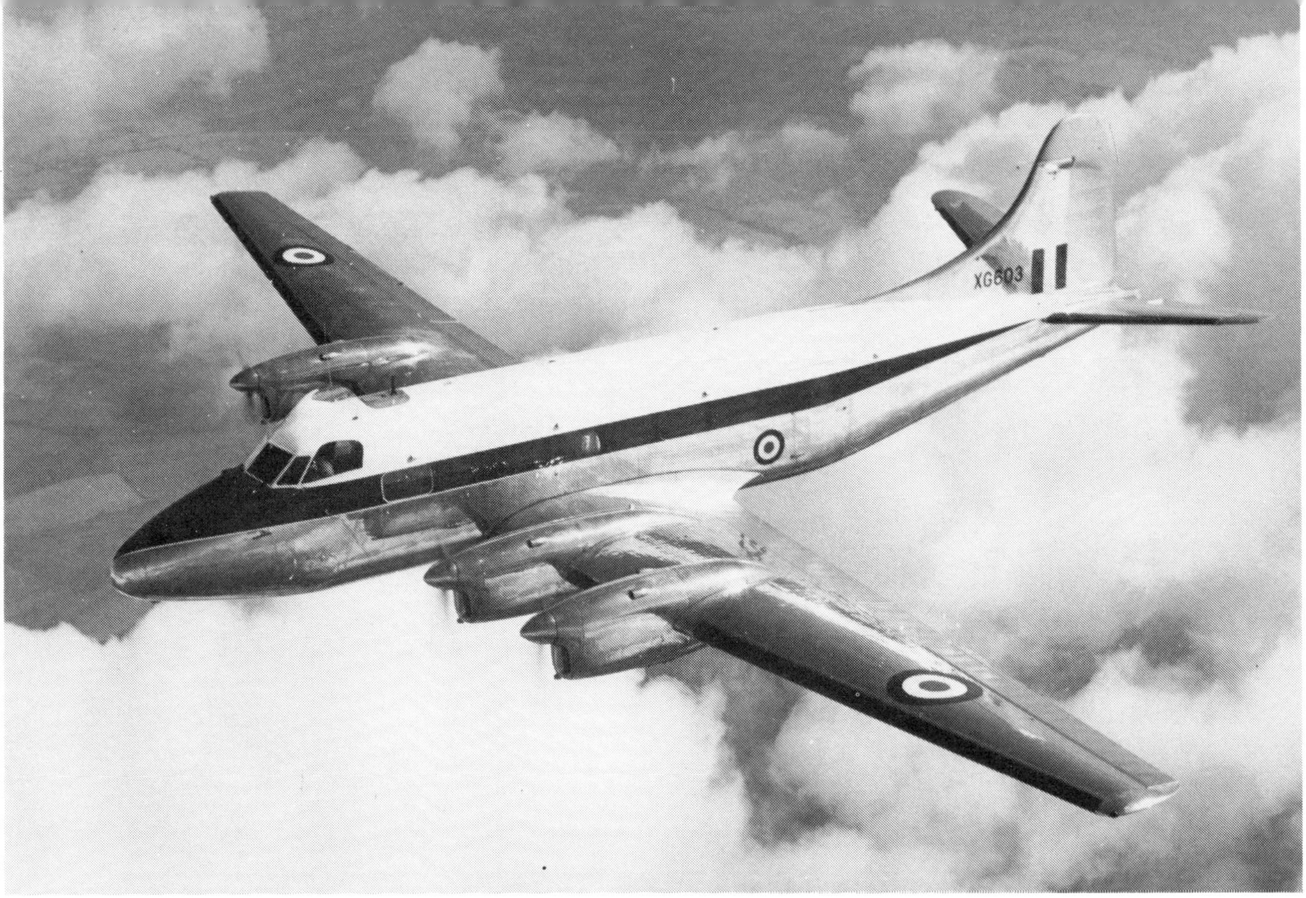
XG603

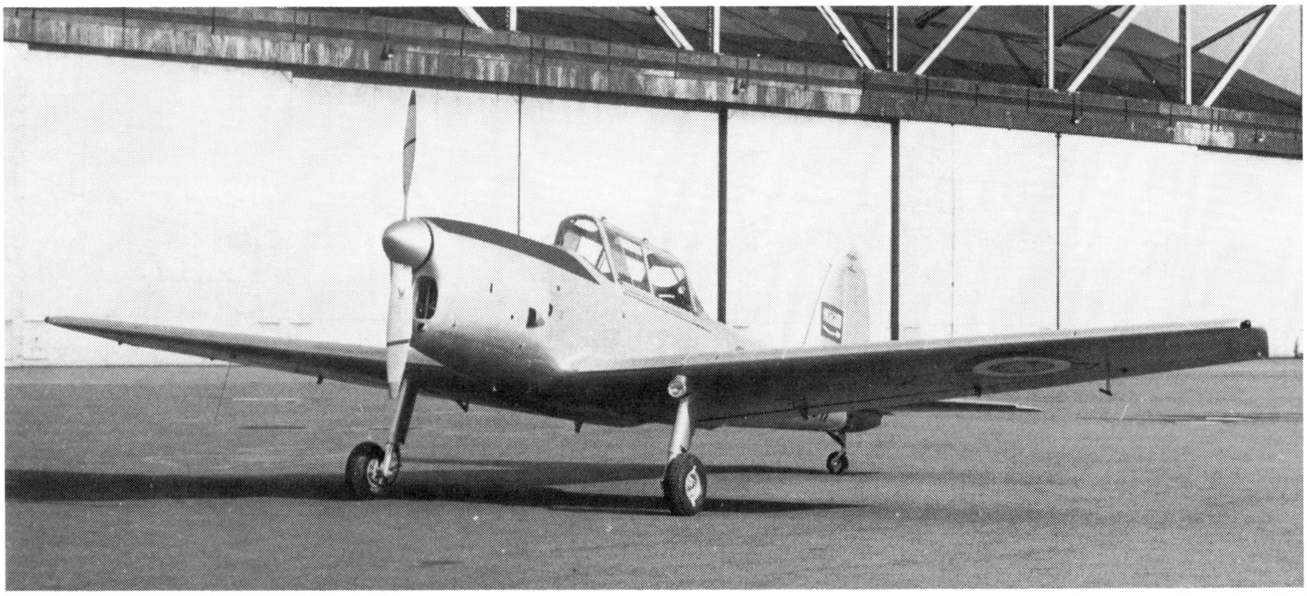

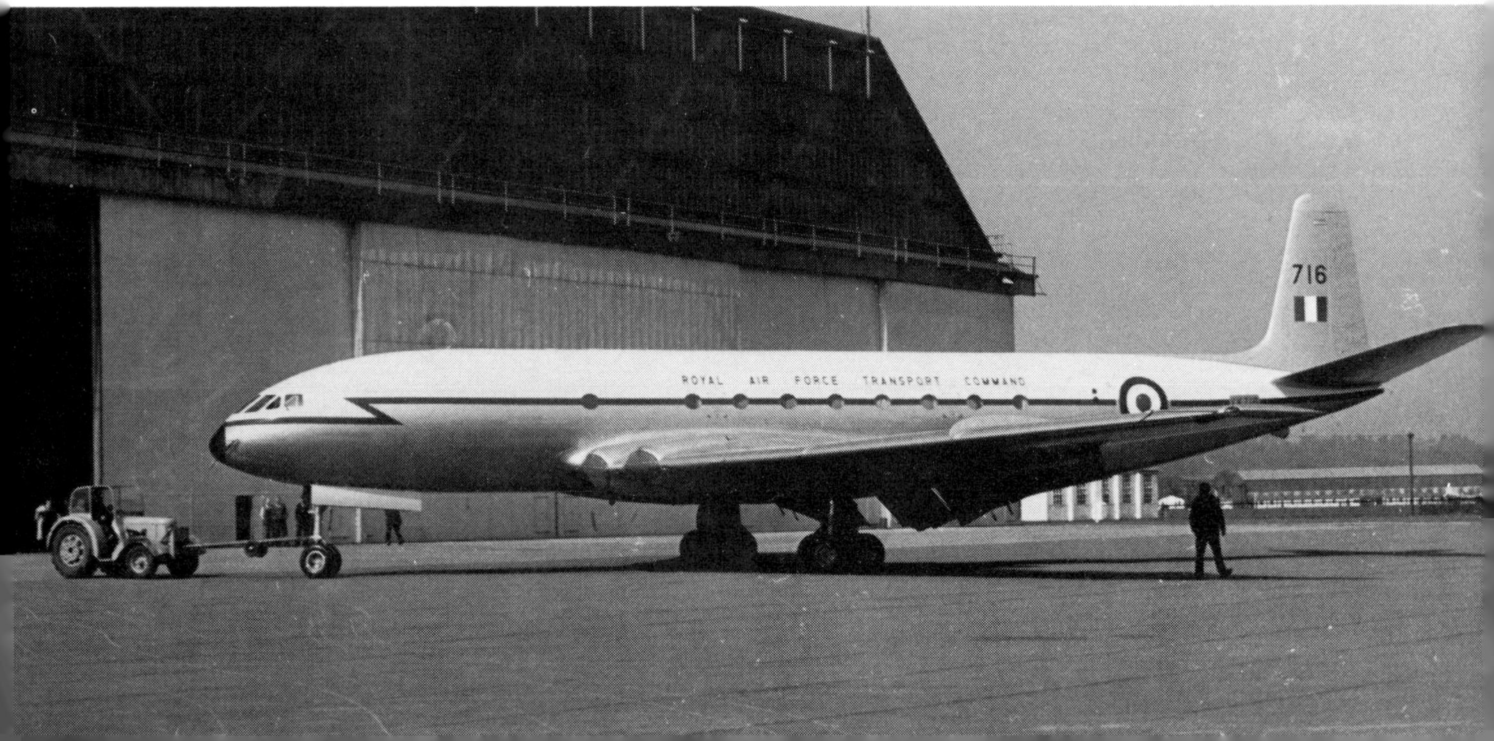
716
ROYAL AIR FORCE TRANSPORT COMMAND

Left Upper:
HERON
D.H.114 Heron, XG603, built at Broughton for the Ministry of Supply seen here during an air-to-air photographic session on 25 August 1954.

Left Centre:
DHC1
An immigrant from de Havilland of Canada, the DHC1 Chipmunk became the RAF's basic trainer to replace Tiger Moths. The 1,000th Chipmunk was recorded on film at Broughton on 14 February 1956.

Left Lower:
COMET
Chester's first Comet 2 (XK716) pictured in the early 1950s.

Above:
DHC2
Broughton-built de Havilland of Canada Beavers G200 and G201 photographed prior to departure on 13 March 1961.

Below:
BROUGHTON
Dating from 1937, the factory was built for 'shadow' production of the Vickers Wellington bomber. Often described as the "Chester Factory", the Broughton works and adjoining Hawarden Airport are now in Clwyd, North Wales.

Scottish Aviation

The big aircraft companies are often considered to be the 'high flyers' of the industry but in 1933 two adventurous Scots flew over Mount Everest and that was high flying indeed.

These two men came from completely different backgrounds; Aristocrat, Douglas Douglas-Hamilton, Lord Clydesdale, was the 14th Duke to carry this hereditary title. He learned to fly in the RAF in which he was a commanding officer of No.602 (City of Glasgow) squadron. His deputy O.C. was David Fowler McIntyre whose family had shipbuilding interests in Scotland.

Seeking publicity, wealthy Lady Houston sponsored the first ever flight over Mount Everest. Two aircraft were acquired for the challenge and the first of these biplanes, a Westland *Wallace* started life as the prototype *Wapiti V (G-ADWA)* but had been rebuilt in 1932 with an enclosed cabin and a Bristol *Pegasus I.S3* engine. Registered *G-ACBR* it was flown over Everest by Flt. Lt. McIntyre on 3 April 1933. On the same day his 'boss' had already achieved the same feat in the *Westland P.V.3* torpedo bomber, *G-ACAZ*, which had been similarly rebuilt and had a *Pegasus I.S3* power unit.

The breakdown of the disarmament talks in Geneva in 1934 signalled the start of a massive build up of armaments in Britain. Large numbers of new aircrew, particularly pilots, were urgently required to make use of the aircraft to be produced by the major manufacturers and the shadow factories. On cost saving grounds the Air Ministry offered contracts to firms willing to set up flying schools.

When Lord Clydesdale and David McIntyre decided to enter the aviation business, as much out of patriotic fervour for Scotland as for capital gain, it was inevitable with their flying background that they should consider that line of activity.

In association with de Havilland which provided the aircraft and some of the capital required a flying school was set up at Prestwick. As the preferred name, Scottish Aviation Limited, was not immediately available for legal reasons the enterprise was established as the Scottish College of Aviation Limited on 9 August 1935.

Flying started at No.12 Elementary Flying Training School (12 EFTS) in February 1936. Twenty orange and silver *Tiger Moths* became a familiar sight around Prestwick in the next few years and to house them the 'Tiger hangar' was erected alongside the grass airfield.

In 1938, training was extended to cover navigation and No.1 Air Observation Navigation School received Avro *Ansons* and Fokker transports which became flying classrooms in a method of instruction pioneered at Prestwick. Later expansion included wireless operator training and all three aspects of Prestwick's initial effort continued until 1941.

The directors of Scottish Aviation were eager to branch out into aircraft manufacturing but at first experienced resistance from the London based Government and wariness from the English companies including de Havilland. Nevertheless, some orders were received to rework Vickers *Wellesleys* at

Opposite:
TWIN PIONEER
G-AOEP pictured against Mt. Kinabalu, North Borneo in the late 1950s. *A. Robertson.*

Above:
FOUNDERS
Group Captain McIntyre (left) and Group Captain the Duke of Hamilton receiving the Freedom of the Burgh of Prestwick on 27 June 1935. *A. Robertson.*

Prestwick. This was supplemented by modification work on Blackburn *Skuas* and *Rocs* which continued until 1941.

For the overhaul of Short *Sunderlands* a repair base was established at Greenock in 1940. Subcontract work from Kingston in the form of rudders for Hawker *Hurricanes* gave Prestwick its first opportunity at making aircraft parts from scratch.

To accommodate all this activity, new buildings were essential and the canny Scots found an excellent 'factory' at a knock down price but with just one small problem that it was in the wrong place. Built in Bellahouston Park, Glasgow in 1938 the 'Palace of Engineering' was part of the Empire Exhibition and was used to show off Scottish wares with a view to gaining orders to boost the ailing local economy. During 1940 the 'Palace' was transported piece by piece to Prestwick where it was re-erected as No.3 Factory. It is still in use today (as No.1 factory).

Until 1943 batches of work came in spasmodically but in that year the pace quickened. On the manufacturing side, sixty-one *Queen Bee* target drones based on the *Tiger Moth* were constructed by SAL in a disused warehouse in Glasgow.

More importantly perhaps, Prestwick itself was chosen as a repair base for fighters, mostly *Spitfires* but some *Hurricanes* as well. In all, 1,200 Spitfires were repaired for return to service and in this SAL was assisted by the LMS Railway workshops at Barassie.

Air training had been established at Prestwick in preference to other locations because of its superb flying conditions. This paid off handsomely when it was selected as a transit base for aircraft sent from the USA on 'Lend-Lease'. From November 1940 until the end of the war, Prestwick was to be kept busy servicing and modifying to European needs the large numbers of overnight arrivals across the 'Air Bridge'. These included heavy bombers such as the B24 *Liberator* and B17 *Flying Fortress* but also C47's, the military version of the DC3 *Dakota*. Similarly, Canadian built *Lancasters* and *Mosquitos* landed at Prestwick on their delivery flights to RAF units.

Below:
WW2
Spitfires, including EP562, with Liberators and a B17 Fortress being repaired in the Palace during the War. *J. Hope.*

Bottom:
SCOTTISH AIRLINES
Liberator II, G-AHZR, (formerly AL552) in Scottish Airlines livery after the War flew as a freighter on BOAC routes. *J. Hope*

When peace returned in 1945 aircraft work was much reduced but Scottish Aviation had a three pronged plan for survival; airport, airline and aircraft.

In the late 'thirties, Scottish Aviation had in mind the establishment of a civil airport to serve both Edinburgh and Glasgow and their hinterland. To double up its flying training the Company had also acquired a site at Grangemouth which was developed as Central Scotland Airport and opened on 1 July 1939. After doing such fine work SAL was rewarded by Grangemouth being requisitioned by the Government as a fighter base and it was eventually bought by the Air Ministry.

Prestwick too was requisitioned in 1941 but, fortunately, SAL was able to hold onto ownership of the site at least for the time being. By 1945 wartime urgency had created a fine modern airfield with large paved runways capable of handling many thousands of arrivals and departures per month. The theme of a prime civil airport for Scotland returned after the war and the Company had many ideas, some grandiose, on how a civil airport should look. Unfortunately it was not alone in its faith for the area and its property was compulsorily purchased by the Ministry of Aviation when it decreed that Prestwick would be Britain's second international airport. SAL had to wait until 1953 for the money; so much for plan A.

Plan B, airline operation, also had its origin in pre-war days and resulted from David McIntyre's belief that active air passenger traffic was essential for a healthy aircraft manufacturing industry. Scottish Airlines was formed in 1946 but nationalisation of the airline routes frustrated attempts to operate scheduled services. Undaunted, charter work was undertaken in addition to the operation of scheduled services under contract to the nationalised companies. Contrary to McIntyre's personal preferences, it transpired that commercial flying was to be a sideline and SAL's main thrust was in aircraft manufacturing and refurbishing.

Scottish aviation turned its wartime experience with American aircraft to advantage by converting war surplus *Liberators* and C.47's for commercial use. With these aircraft, particularly the C.47., SAL acquired a strong clientele and obtained useful funds with a modest outlay.

Above:
DC3
A Sudan Airways DC3 Dakota (front) in company with one of the RAF's C47's being overhauled at Prestwick.

Below:
PIONEER
Prestwick Pioneer in a jungle clearing in the 1950s.
Glasgow Herald

91

Above:
TWIN PIONEER
The RAF acquired Twin Pioneer C.C.1's, including XM286 seen here, to equip squadrons in Africa, Middle East and Far East from 1959.

Scottish Aviation's first complete aircraft, appropriately named *Prestwick Pioneer,* was a rugged single-engined monoplane that would generally be called a bush 'plane. Officially described as a communications aircraft, the prototype *Pioneer, VL515,* with a D.H. *Gipsy Queen* engine flew on 5 November 1947. It was rebuilt with an Alvis *Leonides* engine and temporarily redesignated *Pioneer II*; first flying in this form in June 1950.

Its short take off and landing capability enabled it to operate from places normally considered only possible with helicopters. Forty *Pioneer C.C.1's* were built between 1952 and 1957 for the RAF, and, entering service in 1954, they were utilised in the Malayan jungle in the campaign against terrorists. Their success resulted in sales to the governments of Malaya (9 aircraft) and Ceylon, now Sri-Lanka (4). One *Pioneer* became a film star in "Roots of Heaven".

While the *Pioneer* was being developed, other work was taken in to make use of SAL's metalworking skills. Prominent amongst these tasks was the building of single and double deck coach and 'bus bodies for Albion chassis built in Scotland. A SAL designed light alloy tractor cab was also well received.

Having achieved a modicum of success with the *Pioneer,* SAL proceeded to build a twin-engined derivative. Design work commenced in 1952 and construction started in earnest in 1954 with support from the Ministry of Supply and the Finance Corporation for Industry. G-ANTP, the prototype first flew at Prestwick on 25 June 1955 powered by two Alvis *Leonides 503* piston engines.

The first production *Twin Pioneer* series I with *Leonides 514* engines did not fly until 13 February 1957 but 61 aircraft had been sold by summer 1958. These included 32 *Twin Pioneer C.C.1.* for the RAF which also took on charge 7 *Twin Pioneer C.C.2.*, having long stroke Leonides engines. The *C.C.2.* airframes were similar to the civilian *Twin Pioneer* series 3; the series 2 machines having Pratt & Whitney *Wasp* engines to attract sales from the USA: hopefully but to no avail.

Twin Pioneers sold slower than anticipated and although it was hoped to sell 200 of them the final total was only 87 aircraft but was a good effort for such a small company. The last aircraft was actually sold in 1963. McIntyre was the impetus behind the project and his death on 7 December 1957, when *Twin Pioneer* G-AOEO crashed in the Libyan desert, was a sad loss to SAL and a prime factor in the demise of the *Twin Pioneer.*

The *Twin Pioneer* placed a strain on the Company's financial resources and caused it to diversify once again into aircraft reconditioning in order to survive.

Fortunately, 1955 had marked not only the maiden flight of the *Twin Pioneer* but also the commencement of SAL's association with the Royal Canadian Air Force. In that year, repair and modification of the RCAF's North American *Sabres* had started at Renfrew where it continued until 1960 when work was transferred to Prestwick.

Through its experience with American aircraft generally and contacts in the RCAF in particular, SAL was able to get overhaul work on the Canadair CF.100 and the T.33 trainer. However, a later jet, the CF.104 was by far the most important RCAF aircraft to the fortunes of Prestwick. The first CF.104 arrived on 15 January 1963 and the 1,000th aircraft was signed out before the end of 1967; the contract was to help to keep the works busy for fifteen years all told.

Other American aircraft rejuvenated at Prestwick in the 1960s included the old faithful *Dakota, Skyraiders* for Sweden, DC.8's and 707's.

To oversee this aspect of the business, 'SALcheck' subsidiary was formed in 1962 and the new 'Britannia Hangar' built in 1965 was the temporary home not only for its Bristol namesake but also the *Viscount.*

Prestwick also turned its attention to aero-engines with Pratt & Whitney engines being rebuilt by Scottish Air Engine Services, another subsidiary established in 1960. SAL also became the authorised remanufacturing centre for Rolls Royce *Merlin* and later, *Griffon* engines.

Above:
CAF.104
Royal Canadian Air Force CF.104 fighters being overhauled at Prestwick. The CF.104 was a Canadair-built version of the American Lockheed Starfighter.

Right:
S.A.E.S.
Rebuilding Pratt & Whitney engines at Scottish Air Engine Services, Prestwick, in the 1960s.

Below:
SALchek   SA12
Canadair CL-44 (TF-LLF) owned by Loftleidir of Iceland pictured in the mid-1960s in the Britannia Hangar at Prestwick in company with a DC3 and a Carvair.

Above:
BULLDOG
The second SAL/Beagle Bulldog trainer, G-AXIG, painted with Swedish Air Force markings in 1970.

Right:
BULLFINCH
G-BDOG was actually a series 200 Bulldog which was revamped as a proposed sporting aircraft with retractable undercarriage in 1975.

Scottish Aviation's final decade in the private sector was to be the most important in its history and set Prestwick on a firm foundation for the next twenty years and probably the rest of this century.

In the early 'sixties, SAL had mortgaged its soul to the Finance Corporation for Industry in a desperate effort to survive. However, a chain of unrelated events elsewhere was to make SAL into a fully fledged aircraft manufacturer.

The Finance Corporation (FCI) sold its share in SAL to Cammell Laird of Birkenhead in 1966 which was seeking to diversify from its traditional, but declining, shipbuilding work. SAL now had the backing of an industrial giant.

SAL's 'bread and butter' for the rest of the decade was the recently negotiated contract to supply panels for the Lockheed C.130 Hercules but Prestwick still needed more work. Other people's misfortune became Prestwick's gain in the form of two valuable aircraft types.

Sir Frederick Handley Page's resistance to cooperate with the Government's merger plans was the undoing of the great Company that he had founded in 1909. Even after his death in 1962, his successors continued to defy Whitehall and were denied military orders as a reward. Handley Page designed the H.P.137 twin engined light aircraft to fill a void between the executive jets and the short haul commercial airliners; it was in fact a mini airliner with passenger comfort not usually found in small aircraft.

The prototype, *G-ATXH,* with two Turbomeca *Astazou* turbo-props first flew at H.P.'s Radlett Works on 18 April 1967 and attracted favourable attention and options to purchase from U.S. operators. Prestwick was subcontracted to build 54 sets of wings which required considerably outlay for new plant for chemi-milling and Redux bonding.

Similar heavy expenditure at Radlett overstretched Handley Page Aircraft's resources and it went into liquidation in April 1970. Prestwick lost its wing contract and the rights to the H.P.137 *Jetstream* passed to Bill Bright who then formed Jetstream Aircraft Company.

Shipbuilding was hived off from Cammell Laird which was facing a financial crisis and the residue was renamed the Laird Group in 1970; becoming the new legal owners of SAL.

Beagle Aircraft (formed by (Sir) Peter Masefield after leaving Bristol Aircraft Limited in 1960) built a military trainer version of its *Pup* light aircraft. The prototype, *G-AXEH,* flew on 19 May 1969 but before any further *Bulldogs* could be completed Beagle went into liquidation in 1970. Manufacturing and marketing rights to the *Bulldog* were bought by SAL from the Receiver in 1970 and the first production (SAL) *Bulldog, G-AYWN,* flew on 22 June 1971. Sales were made to Sweden, Kenya and Malaya in addition to 132 aircraft for the RAF delivered from 1973. Altogether, 325 *Bulldogs* were built.

Above:
JETSTREAM T.I.
The second Jetstream trainer XX476 sold for RAF service in 1973 as a Varsity replacement on multi-engine training duties.

Below:
PRESTWICK
The northernmost outpost of the British Aerospace empire (formerly Scottish Aviation) pictured in recent times with the 'Palace' assembly hall prominent in the left-centre portion of this view.

SAL also acquired the rights to the *Jetstream* and proceeded to develop the military version of it which had, in fact, been built and flown as a prototype by Handley Page in 1968. It met the RAF's need for a twin engined pilot trainer and the first *Jetstream T.Mk.1* flew in April 1974. Incomplete airframes built by Scottish and Handley Page which had been stored at Prestwick enabled the RAF order to be filled in record time.

A Garrett engined *Jetstream* was considered by SAL in the mid-'seventies but with the uncertainty over Labour's nationalisation plans it was not built. This project was, however, to prove invaluable to Prestwick's future.

Absorbed Names

As the rationalisation of the aircraft industry took its course after 1960 there were, inevitably, some factory closures.

Percival Aircraft Limited was formed by Edgar Percival and E. Leake circa 1932. Commencing with the *Gull*, it built a series of successful Percival-designed cabin monoplanes at Gravesend, Essex, until a move was made to Luton in 1936. Other single engined machines included the *Mew Gull, Vega Gull, Proctor* and *Prentice* while in twin engine form there were the *'Q', Merganser* and *Prince*.

Percival Aircraft was absorbed by the Canadian owned Hunting Group in 1954 and thereafter traded as Hunting Percival Aircraft Limited until 1957 when 'Percival' was dropped from the title. It built the *Provost, Jet Provost, Pembroke* and *President* until it was absorbed by BAC in 1964.

Gloucestershire Aircraft Company Limited was formed in June 1917 and was owned jointly by the Aircraft Manufacturing Company of Hendon and H.H. Martyn of Cheltenham (an AirCo subcontractor). It acquired the rights to the Nieuport *Nighthawk* and, most importantly, the services of H.P. Folland in 1921.

During 1926, it was re-registered as Gloster Aircraft Company Limited by which time a move had

Above:
PROVOST
The 461st, and last piston engined Hunting Percival P.56 Provost being test flown prior to delivery to the Irish Air Corps.

Below:
JET PROVOST T.51
CJ701 was delivered to the Royal Ceylon (now Sri-Lanka) Air Force in 1959. First flown in 1955, the Hunting Percival P.84 Jet Provost became the RAF's basic jet trainer and lightly armed versions were sold overseas. It was inherited by British Aircraft Corporation when it took over Hunting Aircraft Ltd., in June 1964 and is now built by British Aerospace.

commenced from Sunningend to Hucclecote south of Gloucester.

In the inter-war years Gloster supplied the RAF with its front line fighters powered by Bristol *Jupiter* and later, *Mercury* engines. During the 1939-45 War, Gloster turned out 2,750 *Hurricanes*.

Gloster built Britain's first jet propelled aircraft, the *E28/39* which flew in 1941 and Britain's first operational jet fighter, the twin engined *Meteor*.

Unfortunately, its technical excellence and past achievements did not save it from closure in 1963.

Airspeed Limited was formed in 1931 and built the three engined *Ferry* cabin biplane. Later, as Airspeed (1934) Ltd., it produced the single engined *Courier* and twin motored *Envoy*, a version of the latter being built in vast numbers for the RAF as the *Oxford*. During the War its prime effort was the *Horsa* troop carrying glider and in post War years the *Centaurus* engined *Ambassador* was adopted by British European Airways as its Elizabethan Class short haul airliner.

Airspeed's factories at Christchurch and Portsmouth, Hampshire were taken over by de Havilland in 1954 and were involved with the Sea Vixen and Vampire Trainer until closure in 1962.

Above:
GLOSTER METEOR
WA820, the Meteor F.8 interceptor which established a new record for rate of climb on 31 August 1951. The Meteor was Britain's first operational jet aircraft and about 2,400 had been built in various marks before it was surpassed by the E.E. Lightning.

Below:
AIRSPEED OXFORD
L4576 is representative of the Oxford advanced trainers built from 1938 onward at Christchurch, Hampshire.

BAC

BRITISH AIRCRAFT CORPORATION

BAC

In the ten years following the end of the Second War the British aircraft industry had been reduced in size by about one-fifth due to closures and take-overs. Contraction had slowed in the early 1950s as a result of the war in Korea.

The Conservative party returned to power in 1953 and set about re-organising the affairs of Britain after seven years of socialist reform. By 1956, Britain was in a deep financial crisis so the Government embarked on a series of measures to reduce spending. Defence involves a massive influx of taxpayer's money and therefore did not escape scrutiny.

Plans for change were drawn up in 1956 and published in the Defence White Paper of 1957. This came as a great shock to the industry because it favoured a nuclear arsenal delivered by rockets and was to be enacted purely on the grounds of giving Britain the best possible deterrent at minimum cost, or so it was hoped.

With funds restricted and some embryo projects cancelled outright the prospect for existing aeroplane firms was not good. To make matters worse, it was further declared at meetings with heads of aircraft firms in 1957 that the Government intended to use the remaining military orders to enforce a reduction in size of the industry. Furthermore, the Government was to use its power as paymaster of British Overseas Airways Corporation and British European Airways Corporation to ensure that orders for civil aircraft only went to companies which cooperated. The Government wanted two airframe makers, two engine builders and one helicopter firm.

The top brass of the RAF was not convinced that manned aircraft were a spent force and was also concerned that the use of missiles could hand over power to the other Services. Germany had demonstrated the practicability of army controlled ground to ground ballistic missiles and the German navy had fired the first missile from a submerged submarine during the war. America was developing the *Polaris* submarine launched missile so there was a grave risk that the RAF could be reduced to a subservient role.

Sir Dermot Boyle, Chief of Air Staff, prevailed upon Duncan Sandys' Ministry of Defence for a *Canberra* replacement not as a stop gap until missiles were ready but as a high performance all-purpose machine with a service life of up to twenty years. The result was operational requirement OR339 for which all the major manufacturers submitted schemes for what was to become TSR2 (tactical, strike and reconnaissance).

TSR2 was used as a big stick, some say a shotgun, in Duncan Sandys' plans for reform. A condition of the order for OR339 (later revised to OR343 to freeze out Shorts) was that there was to be an amalgamation of the companies involved.

Having supplied the RAF with its latest and best bomber, the *Canberra*, and the *Lightning* fighter, English Electric at Warton was not prepared to let the TSR2 slip through its fingers. Vickers, however, was equally insistent and the order was given to Vickers in January 1959 provided that it cooperated with English Electric. Vickers was to be the lead site with English Electric as its subcontractor. A design based on the English Electric P.17 delta wing aircraft combining the navigation and weapons system of the Supermarine type 571 after much argument became TSR2.

In 1959 there were thus two major forces in British aviation; the Vickers/English Electric consortium and the firmly established Hawker Siddeley Group. The staunchly independent family firms like Bristol, de Havilland, Blackburn, Fairey, and Handley Page were therefore at risk of being left out in the cold. Consequently, there were discussions between companies with a view to finding suitable partners to appease the Government.

Sir Frederick Handley Page, a very autocratic man, wanted no part of the shotgun wedding plans but Bristol's Sir Reginald Verdon Smith and Geoffrey de Havilland himself were more amenable.

Bristol held discussions in the Hawker camp while de Havilland proposed to join the Vickers/E.E. team. However, while talks were going on the Hawker Siddeley Group in a 'blitzkreig' raid bought the de Havilland concern outright. Hawker Siddeley also took control of Blackburn Aircraft in 1959. This left Bristol out on its own and the only recourse was to join forces with Weybridge and Preston. Handley Page stayed independent and was to perish in 1970 as a result.

A brief mention should be made here of the progress of some of the remaining companies. Shorts in Belfast, Britain's first commercial manufacturer of aircraft, was considered as a special case for political reasons to do with the 'troubles' in Northern Ireland. Westlands had concentrated on helicopters since the war and built Sikorsky designs under licence. In 1960 Westlands took over the helicopter interests of Fairey, Saunders Roe (it had acquired Cierva Autogiro in 1954) and Bristol. Other independents come into the story later.

An announcement on 18 May 1960 giving details of the new British Aircraft Corporation (BAC) concluded many months of hard bargaining. BAC was incorporated on 1 July 1960 as a holding company for Bristol Aircraft Limited (a subsidiary of Bristol Aeroplane Company Limited), Vickers Armstrongs Aircraft Limited (a subsidiary of Vickers Armstrongs) and English Electric Aviation (owned by English Electric Company Limited). Here it should be stressed again that BAC only took over control of the aircraft subsidiaries of these Companies: the main Companies remained independent.

The make-up was 40 percent English Electric, 40 percent Vickers and 20 percent Bristol. BAC also acquired a 70 percent interest in Hunting Aircraft Limited in May 1960 and gained full control in 1963.

On its formation BAC acquired the use of sites at Preston, Warton, Samlesbury, Accrington and Luton

Left:
TSR2
XR219, the only TSR2 to fly, seen here while on a test flight. TSR2 was the keystone to the Tory Government's forced merger plans which created British Aircraft Corporation in 1960.

Above:
BAC ONE-ELEVEN
Phillipine Airlines BAC One-Eleven (PI-C1171) with Rolls Royce Spey turbofans. Based on the Hunting H.107, the BAC One-Eleven was assembled at Hurn (a former Vickers factory) and was the life blood of BAC in the 1960s.

Right Upper:
BAC 221
WG774, the world record breaking Fairey Delta Two as rebuilt at Filton in 1961 with 'ogee' wings and other 'mods' for flight development as part of the BAC (Bristol) type 223 which eventually became Concorde.

Right lower:
HUNTING H.126
The H.126 jet-flap research aircraft which first flew in March 1963 seen here on a test flight from RAE Bedford. Powered by a Bristol-Siddeley Orpheus turbojet it was the last aircraft type to be built by Hunting Aircraft Limited at Luton before it became a part of British Aircraft Corporation in 1964.

owned by English Electric Aviation, Filton and Cardiff ('Bristol'), the Vickers' factories at Weybridge and Hurn and Hunting Aircraft's works at Luton. Preston remained under the control of the main E.E. company but undertook aircraft detail work as well but this was an imperfect situation due to internal politics.

In developing the BAC product range it was agreed that revenue from existing or 'old account' aircraft would go to the companies that had donated the subsidiaries while 'new account' work would accrue to BAC. Vickers, therefore, had chance to recoup its expenditure on the *Vanguard* and VC10 while English Electric retained control of the *Lightning* profits.

With *Britannia* work coming to an end the Bristol airframe factory at Filton was in dire straits. George Edwards, former chief designer at Weybridge and in 1960 chief executive of BAC, ordered that some work on the *Lightning Mk.IV* and VC10 should be transferred to Filton. Conversion of the Fairey *Delta Two* for research with a new ogival wing shape was redirected from Luton to Filton.

Hunting Aircraft at Luton was by far the smallest component of BAC yet its importance surpassed the other three put together in one respect. This was the Hunting H.107 short-haul 50-seat airliner using two Bristol Siddeley BS75 engines. It had competition in the proposed four-engine VC11 and the D.H. *Trident* which although larger had firm sales prospects. Enlarged and with two Rolls Royce RB163 *Spey* engines it evolved as the BAC-III and in May 1961 the BAC Board took the calculated risk of laying down a production run of twenty aircraft. The prototype flew from Hurn in the hands of Jock Bryce on 26 August 1963 by which time orders for the *One-Eleven* exceeded fifty aircraft.

Production was a BAC collaborative effort with Weybridge making the wing skins, fuselage skins and undercarriage; Luton fabricated the wing torsion box; Filton was responsible for the rear fuselage and tail while Hurn, in addition to supplying the front and centre fuselage sections, also undertook the final assembly.

Work sharing was made on the basis of available capacity; Weybridge had already embarked on full scale production of the VC10, Warton and its neighbours had the *Lightning* and Luton had the *Jet Provost* in production. The BAC Board seems to have been reluctant to give too much in the way of long term work to Filton because it wanted a clear deck for the proposed Bristol type 223 supersonic transport and the *Beverley*-replacement Bristol type 208 freighter if they came to fruition.

The *One-Eleven* was a tremendous success by European standards and 222 were to be built up until nationalisation in 1977 and were sold in 63 countries. The success of the *One-Eleven* was fortuitous because without it and existing products BAC might not have lasted longer than five years as subsequent events were to prove.

Mention should be made in passing of the *Belfast* built by Shorts. Originally named *Britannic* and based on Bristol freighter schemes, the Short SC5 *Belfast* used a *Britannia* derived wing built at Filton and was powered by four Rolls Royce *Tynes*. The first one flew in 1962 and the ten to be built went into RAF service from 1966.

To ease the transition to BAC, each constituent site retained the identity of the founding company but on 1 January 1964 a more unified management structure was created. A new subsidiary, British Aircraft Corporation (Operating) Limited took over the assets and businesses of the original subsidiaries. Five divisions were set up under BAC (Operating) Ltd to control the day to day running of Weybridge, Filton, Preston and Luton (Hunting) including the various annexe sites. A Guided Weapons Division formed in 1963 managed Filton and Stevenage guided weapons activity; the Luton factory of English Electric having been closed in 1962 on the cancellation of the *Blue Water* missile system.

Harold Wilson's Labour Party gained control of the House of Commons in 1964, albeit with a majority of only four seats. In the 1965 Budget speech it was announced that TSR2 was to be cancelled and instructions were given for the immediate destruction of all TSR2 aircraft and tooling. Hawker Siddeley also suffered cancellations. Attempts to cancel the SST failed because the French had insisted on a non-cancellation clause in the Memorandum of Understanding.

Above:
BAC 167
NZ361, a BAC 167 Strikemaster Mk.88 sold to the New Zealand Air Force and seen here with extra fuel drop tanks and two Matra rocket launchers inboard. The power plant was a Bristol-Siddeley Viper turbojet.

Orders were then placed for McDonnell *Phantoms* and Lockheed C-130 *Hercules* transports from America. General Dynamics F-111 swing wing bombers were also ordered but never delivered.

Having committed large sums of capital to the *One Eleven*, TSR2 and SST and without the prospect of substantial military or civil orders BAC was in a seriously weakened position. Full advantage was taken of this by the Wilson Government to try to force a merger of BAC with Hawker Siddeley, with BAC as a minority shareholder, but this was strenuously resisted. The threat eventually abated but there were other disruptions to complicate matters still further.

A clash of interests occurred in 1966 between Bristol Siddeley Engines and Rolls Royce. BSE was seriously considering collaboration with Pratt & Whitney of America and SNECMA of France on development of a large turbofan engine (eventually to become the P&W. JT9) for civil use. Rolls Royce at that time was in the doldrums and therefore at risk of being left behind in civil engine sales. To safeguard its position Rolls Royce bought Bristol Aeroplane Company Limited in order to gain control of its subsidiary Bristol Siddeley Engines Limited but in doing so also acquired Bristol's twenty percent stake in BAC which RR did not particularly want.

A similar complication arose in 1968 when GEC bought English Electric Company Limited to gain control of E.E.'s engineering and electronic interests but, like Rolls Royce, also acquired an unwanted share in BAC.

In 1968, BAC was therefore owned by Vickers, GEC and Rolls Royce in the ratio of 40 : 40 : 20. As a result of the new ownership British Aircraft Corporation (Holdings) Limited was formed and the subsidiary BAC (Operating) was renamed BAC Limited. The Divisions remained unchanged.

There had in fact been another change in 1967 when, to consolidate its position, BAC had reluctantly closed the old Hunting factory at Luton. Luton Division was absorbed by Preston Division and all production including the highly successful *Jet Provost* was transferred to Warton.

The first flight of the prototype *Jaguar* in 1968 marked an upturn in the fortunes of BAC. Preston Division had joined forces with Breguet in France in 1966 to form a consortium called SEPECAT to devise a tactical support and ground attack aircraft of high performance. Based on the Breguet Br.121, it was assembled on dual assembly lines at Warton and Toulouse using rear fuselages and wings supplied by Preston Division.

Bristol had its moment of glory on 9 April 1969 when the second prototype BAC/Sud Aviation *Concorde (G-BSST)* made its maiden flight at Filton before going to Fairford Flight Test Centre. Built on duplicate assembly lines at Filton and Toulouse, *Concorde* incorporated parts and sub-assemblies from Weybridge and Preston as well as Filton and French factories.

1971 was another year of turmoil in the aircraft industry. Rolls Royce went bankrupt through its efforts on the RB211 civil turbofan and its twenty percent share in BAC (ex Bristol) was bought from the Receiver by GEC and Vickers which then each held a half share of BAC.

Other re-organisation in 1971 resulted in a new Military Aircraft Division which included Preston, Warton and Samlesbury; a Commercial Aircraft Division at Filton and Weybridge; and a Guided Weapons Division to encompass G.W. activity at Filton and Stevenage.

For the next few years BAC was untroubled, relatively speaking, and could get on with building

Jaguars, Concordes and *One-Elevens* in an attempt to sustain its profitability.

In 1974 the product range was expanded by the introduction of the *Tornado* swing wing aircraft; a culmination of thirty years indirect effort. It is believed that the Germans were the first to experiment with variable geometry wings in the Second War. In Britain, Barnes Wallis at Vickers was the chief exponent of pivoted wings and built rocket propelled models under a Ministry experimental contract in the late 1940s and early 'fifties. Full size wing pivot test rigs were built at Weybridge but all work was transferred to Warton in 1964.

Proposals to build an Anglo-French variable geometry aircraft (AFVG) fell on stoney ground due to political problems. Warton, however, persisted and in conjunction with MBB of Germany and Fiat in Italy a consortium (PANAVIA) was formed in 1969 to build the multi role combat aircraft (MRCA).

Final assembly lines in each of the three countries were responsible for the aircraft destined for its respective air force but there was no duplication of part and sub-assembly manufacture. BAC for its part supplied all the rear fuselages, nose sections and cockpits while MBB was given the task of making the centre fuselage and wing pivot. Aeritalia, formerly FIAT, designed and built the wings.

State of the art technology enabled one basic *Tornado* airframe to fulfill six major operational roles, day or night in all weather at high altitude supersonic speeds or low level subsonic flight. The first version to emerge was the interdictor strike (IDS) which was joined later by an air defence variant (ADV).

Being a consortium, PANAVIA was not greatly affected by the corporate transition from BAC to British Aerospace.

Time has proved that the BAC Board chose the right long term corporate strategy even if the short term financial gain was relatively small. As a result, BAC had much to offer the new nationalised corporation and the effects are still evident today.

Below:
CONCORDE
Concordes in final assembly at Filton in the late 1970s with -212 (G-BOAE) in the foreground. Concorde is still the world's only operational supersonic transport and the only aircraft in the world that can fly at Mach 2 for extended periods in daily service.

Top:
BLUE WATER
A tactical weapon for the Army, the Blue Water missile system was a technical success. Its cancellation on cost cutting grounds was a severe blow to BAC and caused the closure of Luton (ex English Electric) factory.

Above:
VIGILANT
Vickers first generation anti-tank missile equipped the British Army until superseded by Swingfire.

Below:
RAPIER
The Rapier low-level anti-aircraft system was designed and built at Stevenage to compliment the high altitude Bloodhound and Thunderbird surface to air missiles.

Above:
ARIEL 4
Ariel 4 during integration at Bristol. BAC was the prime contractor for this scientific satellite designed for ionospheric measurements.

Below:
SKYLARK
The transfer by the Government of responsibility for Skylark from Royal Aircraft Establishments to BAC brought work to Filton and Banwell (Bristol Aerojet) in the 1960s.

Above:
JAGUAR
Originally the result of cooperation between BAC Warton and Breguet, the Jaguar ground attack aircraft was built by an Anglo-French consortium known as S.E.P.E.C.A.T.

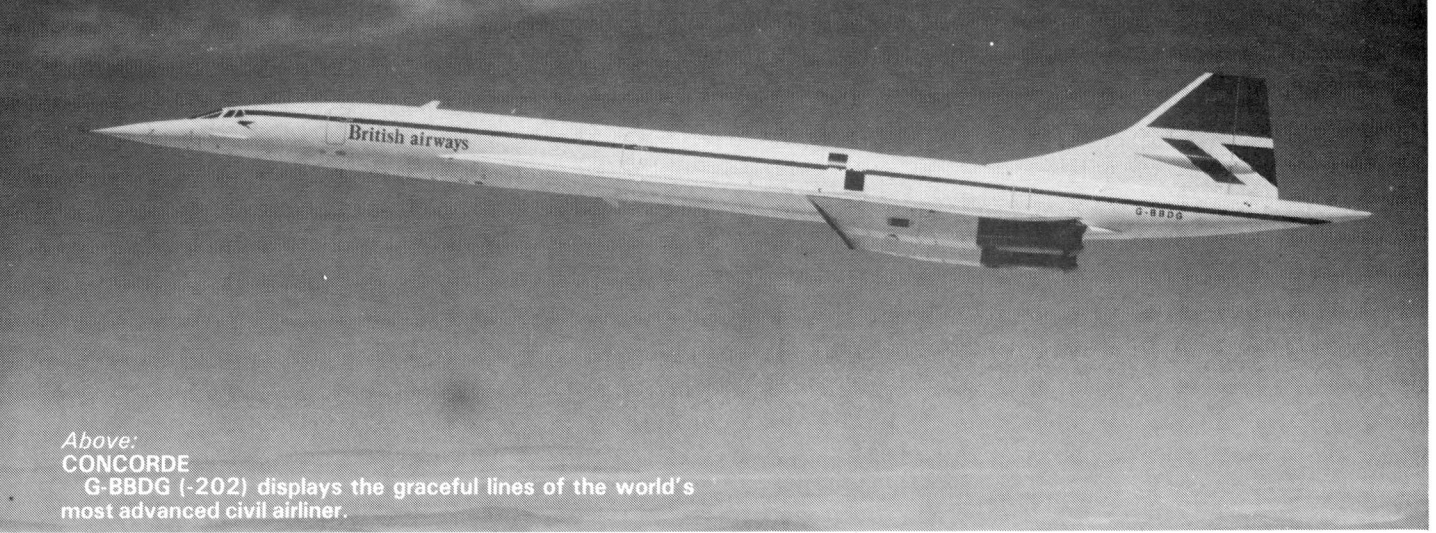

Above:
CONCORDE
G-BBDG (-202) displays the graceful lines of the world's most advanced civil airliner.

Below:
PANAVIA TORNADO
The second of nine prototypes, XX946 was the first British-built Tornado and first flew at Warton on 30 October 1974.

HAWKER SIDDELEY

Even before the 1957 Defence White Paper was published, Hawker Siddeley had developed into a very powerful and influential group of aircraft manufacturers. Before unravelling its progress from 1960 onward it would be wise to amplify some of its past history.

Although the Company was partly named after Harry Hawker who met his untimely death in 1921, it was Thomas Sopwith (he was knighted in 1953) and Sir John Siddeley who were the architects of the enterprise. T.O.M. Sopwith's involvement in British aviation has already been elaborated in a previous chapter but the role played by his contemporary should now be explained.

In 1902, Mr. J.D. Siddeley formed the Siddeley Autocar Company to manufacture 6, 12 and 18hp cars sold at prices starting at £175 for the 6hp model.

From 1905 to 1911, John Siddeley was associated with the Wolseley Tool & Motor Car Company which produced a range of Wolseley-Siddeley cars.

In 1911 he severed his contact with the Wolseley Company to join forces with Captain Deasy in the formation of the Siddeley-Deasy Motor Car Company. Apart from turning out an extensive range of cars, this concern built the six-cylinder water-cooled *Puma* aero-engine at a rate of up to 700 engines per month in the 1914-18 war. *Puma* engines powered many famous World War One aircraft including the *Bristol Fighter*, D.H.9., and the original Avro *Manchester*.

Discussions in 1919 between Siddeley Deasy and Sir W.G. Armstrong Whitworth & Co. Ltd., led to the formation of a holding company, eventually to be known as the Armstrong Siddeley Development Company Limited. This controlled two subsidiary companies called Armstrong Siddeley Motors Limited and Sir W.G. Armstrong Whitworth Aircraft Company Limited.

The first mentioned company commenced production of air cooled radial aero-engines in 1922 of which the best remembered are the *Jaguar, Genet, Lynx, Tiger* and *Cheetah*. Armstrong Siddeley Motors also built luxury cars. Armstrong Whitworth must take the credit for the RAF's first all metal fighter, the *Siskin* which entered service in 1923, and its army cooperation derivative named *Atlas* which flew in 1924. Another subsidiary of Armstrong Siddeley Development Company was High Duty Alloys Limited, makers of light alloy forgings, stampings and extrusions.

The success of the *Siskin, Atlas* and aero engines helped to lay the foundation for further expansion and in 1928 Armstrong Siddeley Development acquired A.V. Roe & Co. Ltd. By 1935, Sir John Siddeley not only controlled these companies but also Air Service Training Limited and several others.

In 1935 Hawker formed a trust to acquire the Armstrong Siddeley Development Company and a new holding company was set up known as Hawker Siddeley Aircraft Company Limited. The combination of these two famous names was a prelude to an expansion of the aircraft and engineering interests by

Opposite:
TRIDENT THREE
G-AWYZ pictured here on 19 March 1970, four months after its maiden flight, was the first of 26 HS 121 Trident 3B-101s sold to BEA in the 1970s.

Below:
SIDDELEY SISKIN
The wooden framed Siddeley S.R.2 built by the Siddeley Deasy Motor Car Company at Coventry in 1919 and powered by an ABC Dragonfly engine.
Siskin III, the production version of 1923, was the first all metal fighter to enter RAF service.

the taking over of many other established companies. Sir John Siddeley retired from business after the merger and was later to join the House of Lords as Lord Kenilworth.

Hawker Siddeley Aircraft played its part in re-equipping the RAF for war, with everything from interceptor fighters to heavy bombers and built 40,000 *Cheetah* engines during the war. Besides the main factories at Kingston, Langley, Hamble, Gloucester, Coventry, Manchester and Yeadon, it had dispersal sites in at least forty locations ranging from Sedbergh in the north to Paignton in the south.

It was in 1948 that Hawker Siddeley Aircraft Company was renamed Hawker Siddeley Group which through Hawker Siddeley Development Company controlled twenty-five wholly owned subsidiaries by 1951; seven of which were principally involved in airframe or aero-engine manufacture.

By 1960, Hawker Siddeley conformed to the Government's ideal of a large powerful conglomerate able to resist overseas competition and was well able to benefit from Duncan Sandy's plans.

The take-over of the Blackburn Group, Folland Aircraft Limited and the de Havilland group of companies (then comprising airframe, aero-engine and propeller companies in Britain, Canada, USA, Australia, New Zealand and South Africa) added three more arrows to the Hawker Siddeley bow and made it a force to be reckoned with.

For the next few years the individual companies retained their identity and traded very much as before but on 1 July 1963 a reorganisation of the Hawker Siddeley Group gave birth to two new 'principal companies'; Hawker Siddeley Aviation Limited and Hawker Siddeley Dynamics Holdings Limited.

Under HSA, three major aircraft units were created; Avro Whitworth Division (combining Avro and Whitworth-Gloster), Hawker Blackburn Division (including Folland at Hamble) and the de Havilland Division.

Hawker Siddeley Dynamics Holdings Limited was not split into divisions but into three separate companies, Hawker Siddeley Dynamics Limited; builders of guided weapons, mechanical equipment such as undercarriages, propellers and space systems. Hawker Siddeley Dynamics Engineering Limited undertook the manufacture of aircraft equipment, marine equipment, controls for aircraft, controls for aero-engines and industrial processes together with digital test equipment and automatic welding. S.G. Brown Limited was a small company specialising in gyroscopes, navigation and communications equipment.

At the time of the reshaping, Hawker Siddeley's aero-engine business had been amalgamated with that of Bristol Aeroplane Company. Bristol Siddeley Engines Limited, formed in 1958 by merging Bristol Aero Engines and Armstrong Siddeley Motors, took over de Havilland Engines and Blackburn Engines Limited in 1961.

Through its diversity of interests Hawker Siddeley was in a stronger financial position than BAC which was dependent upon a few very large politically vulnerable projects.

Inevitably, the types of aircraft being built by Hawker Siddeley Aviation owed their existence to the founder companies and the first true HSA aircraft was still a decade into the future.

In the Avro Whitworth Division, Manchester was seeing the run down of the *Vulcan* line but was pinning its hopes on the Avro 748, renamed HS 748. Following a worldwide sales tour of the Series-2 prototype and demonstrator, *G-ARAY*, in the spring of 1963, orders were received in the next few years from operators in Brazil, Bahamas, Philippines as well as in Britain.

The HS 780 *Andover C.1* (nee Avro 780), a military development of the civil HS 748, made its first take-off on 9 July 1965 and the thirty examples sold to the RAF were a welcome boost to the Chadderton detail shops and the Woodford assembly lines.

Right Upper:
HS748 COASTGUARDER
Custom built for fishery protection duties, pollution control and off-shore patrol; the Coastguarder could also carry liferafts for air sea rescue work.

Right Lower:
P.1127 KESTREL
One of the 'Tripartite Nine', XS688 was flown in the 1960s by British, German and American crews in exercises sponsored by the U.S. Mutual Weapons Development Fund.

Below:
ARGOSY
G-APRN, the third HS 650 Argosy appeared at Farnborough in 1959 and served as a demonstrator in Europe and the USA until sold to BEA in 1962.

Baginton (Coventry), and nearby Bitteswell aerodrome were at work on the *Argosy* which was unfortunately to be the end of the runway for the former Armstrong Whitworth factories. The HS 650 (ex A.W.650) *Argosy* had found favour with civil operators in Britain and the USA and ten were sold. Fifty-six of a military variant, HS 660 *Argosy C.Mk.1* were delivered to the RAF commencing in 1962 and had 'clam shell' rear doors for supply dropping and paratrooping. These were followed by eight of the larger bodied *Argosy* 200 series, the first of which, *G-ASKZ* flew at Bitteswell on 9 March 1964. Original *Argosies* had wings designed and built in Manchester and tails from Gloster but later versions had a redesigned wing.

Combat aircraft division would have been an equally appropriate title for the Kingston Brough Division of HSA which had three jet aircraft; *Buccaneer, Harrier* and *Gnat* on the jigs at Brough, Kingston and Hamble respectively.

Production of the *Buccaneer S.1* having two B.S. *Gyron Junior* engines ended at Brough in December 1963 by which time the Rolls Royce *Spey* engined *Buccaneer S.2* had flown. *Buccaneer S.2's* for the Fleet Air Arm were built until 1968 and fifty-four were eventually delivered.

Meanwhile, at Kingston H.Q., the *Harrier* close support and ground attack fighter was taking shape. Owing to political factors and in view of the novel features of its Bristol Siddeley *Pegasus* power plant the *Harrier* had a fairly long development phase and the first of six pre-production machines *(XV276)* did not fly until 31 August 1966: nearly five years after the first flight of the prototype P.1127. After the flight of the first production *Harrier (XV738)* in December 1967 events moved more quickly so that *Harrier G.R. Mk.I's* entered RAF service with No.1 squadron, Wittering, in April 1969 and were deployed in Germany one year later.

Hamble's contribution, the *Gnat,* was being built in the trainer version in the early 'sixties to equip RAF Flying Training Schools where they were to give more experienced pupils their first taste of flight at transonic speeds. *Gnat Trainers* were fully aerobatic and will be best remembered for their performance with the 'Red Arrows' aerobatic team.

Gnat production ceased in mid-1965 and thereafter Hamble was assigned to the making of details and sub assemblies for the *Harrier* and *Trident*. Dunsfold, incidentally, was the final assembly point and flight test airfield for *Harriers* and *Gnats* which were delivered in sections by road from the respective sources.

The importance of de Havilland was reflected in it being made a division in its own right. Hatfield and Chester adopted Hawker Siddeley policies but overseas, where the de Havilland name was well known to customers, the transition occurred more slowly.

Although they were essentially de Havilland designs, the D.H.121 *Trident* and D.H.125 business jet were developed into a commercial success in Hawker Siddeley days.

The *Spey* engined HS 121 *Trident* first flew at Hatfield on 9 January 1962 and BEA, which had specified its requirement for it in 1956, took delivery of 23 *Trident IC.'s* from 1964. Similar 115-passenger *Trident IE.'s* were sold to airlines in Ceylon, Cyprus, Iraq and Kuwait.

A longer range variant, *Trident 2E*, having the 'stretched' fuselage of the *Trident IF.*, increased wingspan and extra fuel capacity was tailored to suit BEA's Middle East routes. Fifteen of this 97-seat variant went into BEA service from 1965.

Small as the HS 125 was, it had an importance to HSA (and later BAe) that was great in proportion to its size.

After the first flight of the two prototypes at Hatfield in 1962, the first production aircraft, G-ARYC, flew at Chester on 12 February 1963. Chester then took on full responsibilty for HS 125 production and an assembly line was established at Hawarden, sometimes called Broughton.

Just as the *Devon* had been a military version of the *Dove*, so the *Dominie* was a militarised HS 125. The first of twenty HS 125 series 2 *Dominie T.Mk.I* navigation trainers for the RAF was flown at Chester on 30 December 1964 and these were later supplemented by four V.I.P. courier transports.

HSA had its setbacks in the mid-'sixties. Products such as the *Gnat* and *Argosy* had reached the end of their production life by that date and new projects were needed. The key developments were the P.1154 designed at Kingston as a supersonic V/STOL fighter with a Bristol Siddeley BS.100 engine and the HS 681 proposed by Coventry as a V/STOL transport using four Bristol *Pegasus* engines. It was proposed that the HS 681 and P.1154, would form a complimentary unit with the HS 681 providing logistic support in a fast moving battle front using quickly prepared or naturally available landing sites.

As well as cancelling the BAC TSR2 in 1965, the Wilson Government had also nipped the P.1154 and HS 681 in the bud in 1964.

Diversity was the strength of Hawker Siddeley but in some respects it was a weakness for, having acquired so many aircraft sites, there was bound to be some duplication of effort and over capacity. The Gloster factories which closed in 1963 were the first

Above:
DOMINIE
XS709, the first HS125 Dominie T.Mk.1 navigation trainer for the RAF made its maiden flight at Broughton in December 1964. Power was supplied by two Bristol Siddeley Viper 520 engines.

casualty and with the cancellation of the HS 681, Baginton was to be next but Bitteswell survived until after nationalisation.

Some new work was, however, forthcoming in the late 1960s besides improved versions of the existing range. Notable amongst these 'new' types was the HS 801 *Nimrod* which was, in fact, a major rebuild of the D.H.106 *Comet 4C* as a replacement for the ageing *Shackletons* in the maritime reconnaissance role. The prototype conversions XV148 and XV147 were completed in 1967 at Chester and Woodford respectively. Rebuilding involved adding a secondary fuselage below the main pressurised shell to store depth charges and homing torpedos. *Nimrod MR.I* and *MR.2* aircraft have equipped RAF Strike Command squadrons since 1970.

New generation aircraft had occupied the thoughts of the HSA design offices in the 'sixties.

Designed as an advanced jet trainer with worldwide appeal, the HS 1182 (numbered in the Kingston project series) and later named *Hawk,* was the winner of an official 'competition'. Intended to replace three RAF types — *Jet Provost* for basic training, the *Gnat* for advanced flying instruction and the *Hunter* for weapons training, the *Hawk* concept also included provision for its use in certain combat roles to reduce the defence costs of small air forces.

Such was the confidence in the *Hawk* concept that it was ordered straight off the drawing board without the usual prototypes and pre-production: early production Hawks being allocated for development. The first *Hawk T.Mk.I* (XX154) flew at Dunsfold on 21 August 1974 and an initial order for 175 machines guaranteed its commercial success. *Hawks* went into RAF service in 1976.

Right:
BLUE STREAK
De Havilland's Blue Streak intercontinental ballistic missile was the principal weapon of Britain's nuclear deterrent until it was realised that static launch sites would be vulnerable to counter-attack. It was thus cancelled in 1960 but was promoted by Hawker Siddeley as the first stage of a three-stage European vehicle for satellite launching.

Below:
NIMROD M.R.1.
Prototype maritime reconnaissance Nimrod, XV148, at Broughton. Converted from a Comet with R.R. Spey engines it first flew on 23 May 1967.

Above:
HS1182 HAWK
First production Hawk T.I trainer of 1975 vintage (XX156) with Rolls Royce/Turbomeca Adour engine.

Below:
HARRIER GR.3.
XV778 is shown here during exercises in Norway.

Right Upper:
TRIDENT 3B
G-BAJM, a Trident 3B-104, was first registered in November 1972 and sold to the Peoples Republic of China in October 1975.

Right Lower:
BH125
Beechcraft Hawker 600 Jet demonstrator appropriately registered N125BH.

Meanwhile, Hatfield's design office had, during the 1960s, sought ways of bridging the gap in the product range for a short haul feeder liner with short take off and landing capability. Of the many schemes prepared during this time the HS136 (1964) was a low wing aircraft of conventional appearance with two engines in underwing pods. In contrast, the HS 144 (1970) had two turbofans mounted on the body just aft of the wings and had a 'T' — tail.

From these and other ideas evolved the high wing HS 146 having four, small, fuel-efficient turbofans in wing mounted pods. By 1974 work had started but a recession in industry worldwide together with the oil crisis which saw oil prices quadruple and the looming threat of nationalisation caused the project to be shelved.

In the last five years of the existence of HSA prior to

nationalisation in 1977, its various factories were sustained by a mixture of revamped old types until production of the new types got into full swing.

The phasing out of aircraft carriers to save money spoiled chances of further Naval orders for Brough but when the American F-III ran into technical problems and the AFVG floundered at the design stage, *Buccaneers* were adopted by the RAF. Conversion of withdrawn Royal Navy *Buccaneers* and later new production for the RAF sustained Brough works until the end of 1976.

At Kingston, Dunsfold and Hamble the *Harrier* was the prime source of work until *Hawk* production gained momentum. Hamble also had some sub-contract work from Hatfield in the form of details for the *Trident 3B*.

For BEA's medium range, high density routes, the *Trident 3B* had a 16ft 3in longer fuselage than the *Trident IC* and more powerful *Spey* engines. BEA bought 26 *Trident 3's* for inter-city services and the first of these went into revenue operation in 1970.

The Hatfield assembled *Trident* pioneered the civil use of fully automatic 'blind' landing and G-ARPB, a *Trident IC.*, achieved distinction by landing 'hands off' at RAE Bedford in March 1964. Fog is one of the biggest nightmares of airport operation and G-ARPR made flying safer by being the first civil aircraft in the world to make an automatic landing in fog when it touched down at Heathrow in November 1966.

At Chester, the extra thrust available from the Bristol Siddeley *Viper 522* engine enabled the gross weight and maximum speed of the HS 125 to be increased. Designated 'series 3', over half of the 53 to be built were exported including to places as far away as Australia.

Penetration of the American market, started with the HS 125 series 1, continued, and in order to become more deeply entrenched Hawker Siddeley signed a collaborative agreement with Beech in 1969 to market the BH.125 in the USA by the Beechcraft Hawker Aircraft Corporation. This concord was later terminated by mutual agreement.

At this juncture, the series 400 was introduced which featured more luxurious seating, an improved flight deck and had a folding stairway incorporated into the passenger door.

The flight of G-AXYE at Chester in January 1971 marked the debut of the HS 125 series 600 with *Viper 601* engines; the form in which the small jet was to be produced from 1972 until nationalisation.

BRITISH AEROSPACE

By the late-'seventies, British aviation had witnessed many changes and upheavals. The volatile nature of the aircraft market shaped the destiny of countless thousands of people of many persuasions who earned their living through involvement in aircraft work. Under pressure from market forces and political doctrine the stage was set for another transformation scene.

On 29 April 1977, under the provisions of the Aircraft and Shipbuilding Industries Act 1977, the issued share capital of British Aircraft Corporation (Holdings) Limited, Hawker Siddeley Aviation Limited, Hawker Siddeley Dynamics (Holdings) Limited and Scottish Aviation Limited was vested in a new nationalised corporation called British Aerospace. The assets and businesses of these four companies were acquired by British Aerospace on 1 January 1978.

To recognise the two main branches of the enterprise an Aircraft Group and a Dynamics Group were formed which were in turn split into six and two divisions respectively.

The status quo was short-lived for on 1 January 1981 the structure of the Company was changed by the British Government from a corporation under national ownership to a public limited company in the private sector. One hundred million shares representing half the issued share capital were sold at that time.

MILITARY

On the military front, four combat aircraft stayed in production for the entire span of BAe's first decade; two from each of the predecessor companies.

The Interdictor Strike (IDS) version of the PANAVIA *Tornado* went into full production to equip the air arms of Britain, West Germany and Italy; the first production aircraft being delivered in 1980. The prototype *Tornado ADV* (Air Defence Variant) first flew on 27 October 1979 and the first production order for the RAF was signed in August 1982. Over one thousand *Tornados* are expected to be built including 72 for the Royal Saudi Air Force.

Another collaborative venture, *Jaguar International,* the export version of the SEPECAT *Jaguar* low level supersonic strike aircraft, achieved sales to Oman, Ecuador, Nigeria, and India where it is being built under licence by Hindustan Aircraft.

A new variant of the unique *Harrier,* the world's only successful V/STOL combat aircraft, evolved as a result of experience with the US Marine Corps' *AV8As.* The British Aerospace/McDonnell-Douglas Harrier II (known in the USMC as *AV8B* and as *GR5* in the RAF) features a carbon fibre composite wing and other improvements to give it increased range and greater payload than previous versions.

On the ground, the British 'invention' of the *Ski-Jump* enabled the *Harrier* to take-off in a shorter deck run or with higher gross weight. *Sea Harriers* are operational with the Royal Navy on *Ski-Jump* equipped *Invincible-Class* "command" cruisers.

The *Hawk* has made a valuable contribution to Britain's export earnings and on completion of the RAF's order for 175 *Hawk T.Mk.1's* sales of *Hawk Mk.50 Series* aircraft were made to Finland, Africa and Indonesia. Faster and more versatile *Hawk Mk.60's* were sold to Zimbabwe and middle east nations including Kuwait. Further developments include the *Hawk 100* advanced ground attack aircraft and *Hawk 200* single seat fighter.

The *EAP* (Experimental Aircraft Programme) demonstrator, the first all-new aircraft to be assembled by BAe, was built with European help to test active fly by wire technology in an aerodynamically unstable aircraft. Since its first flight on 8 August 1986, it has been used for work to support the *EFA* (European Fighter Aircraft) which, it is proposed, will be built as a collaborative venture by 'Eurofighter' to compliment the *Tornado* for service in the 1990s.

Left:
TORNADO ADV
The initial Air Defence Variant of the PANAVIA Tornado is known in the RAF as the F.2 and has two Turbo-Union RB199 Mk.103 engines. Tornado F.3's have more powerful RB199 Mk.104's. Delivery of the RAF's 165 aircraft commenced in 1984.

Below:
JAGUAR INTERNATIONAL
XX108 demonstrated at Paris in May 1979 the ability of the export version of the SEPECAT Jaguar to land on grass airstrips. Delivery of this model to the Nigerian Air Force began in 1984.

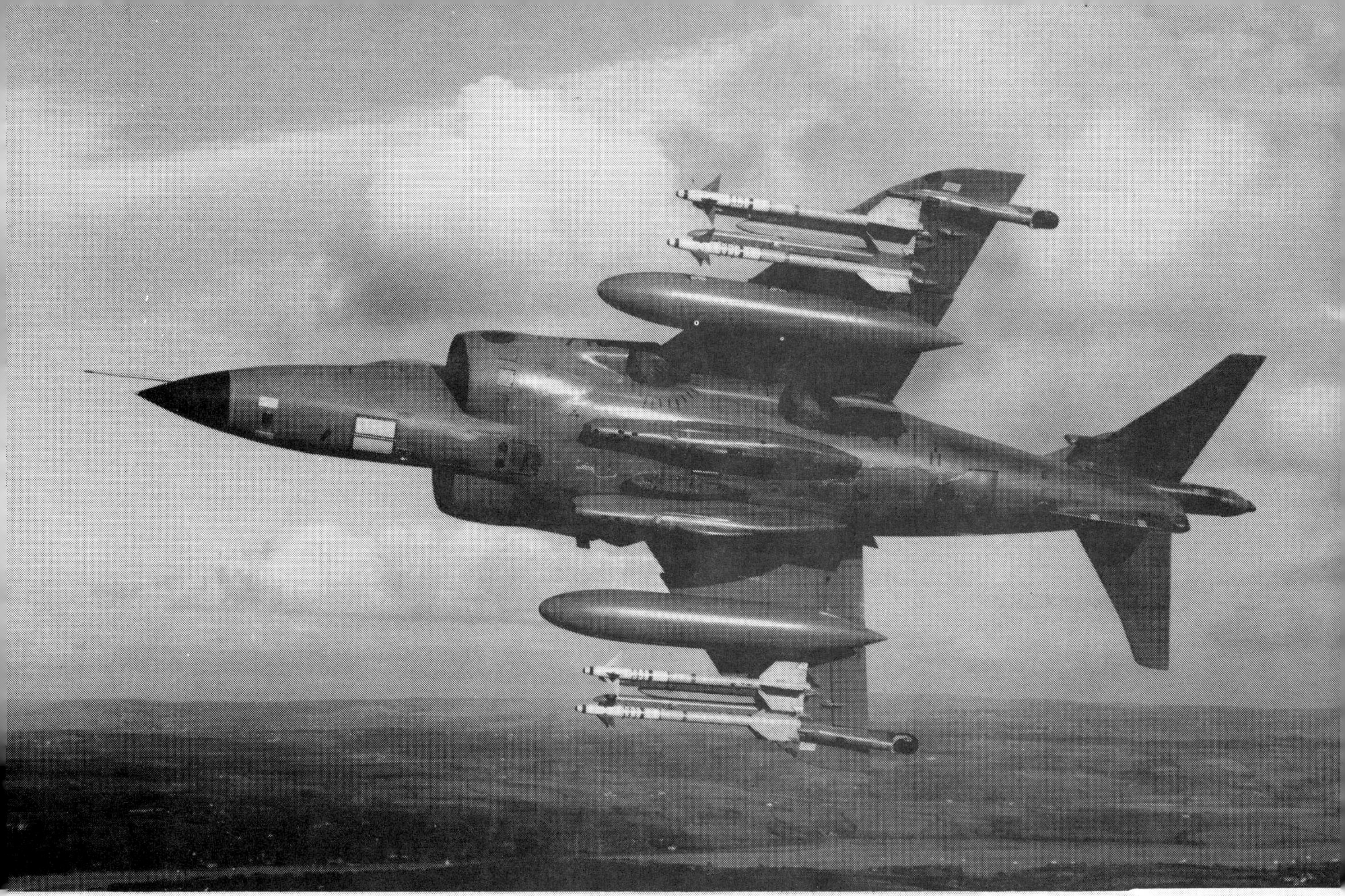

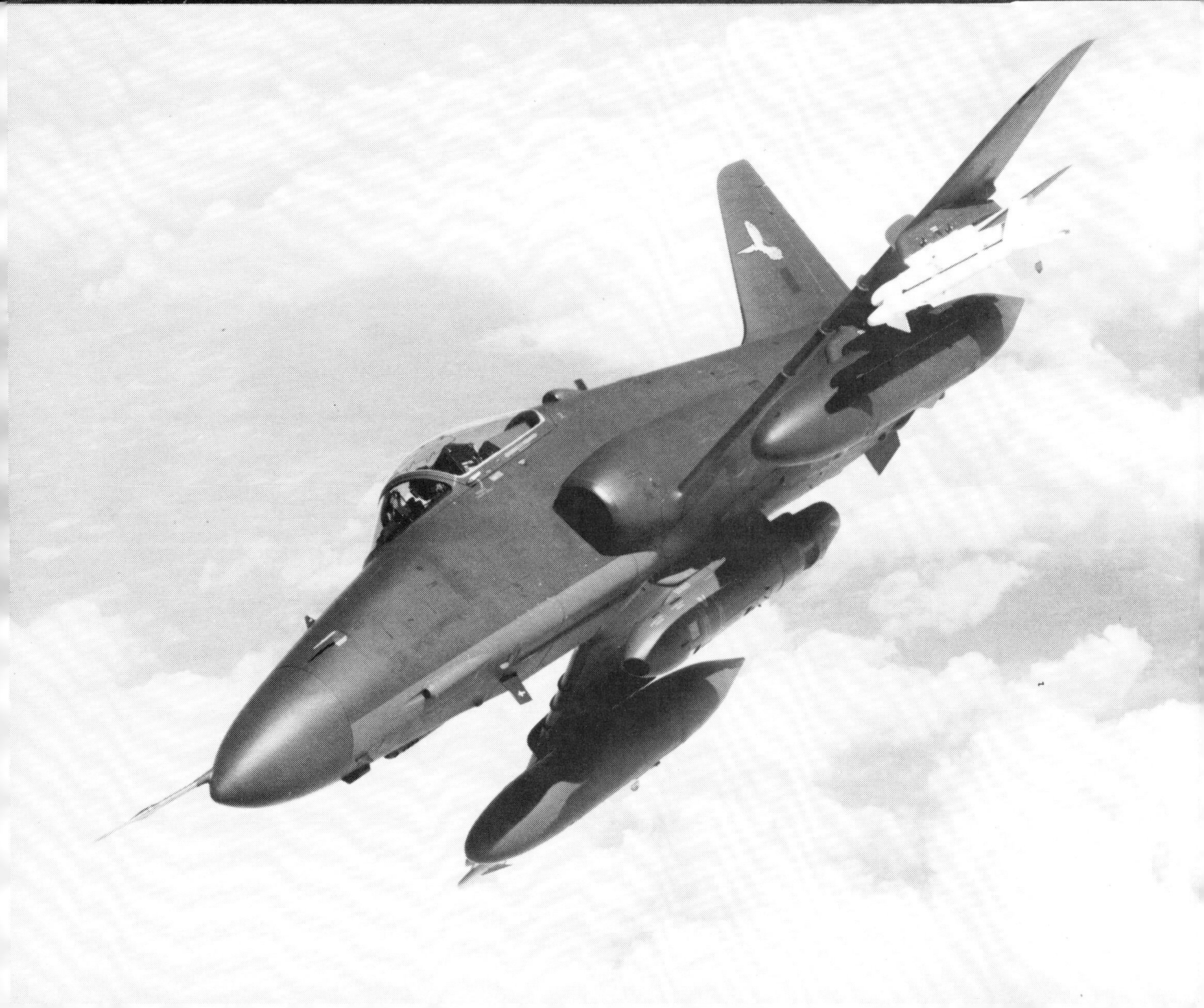

Left Upper:
SEA HARRIER
   XZ457, a Sea Harrier F.R.S.1 with Rolls Royce Bristol Pegasus engine. Sea Harriers saw active service in the Falklands campaign in 1983.

Left Lower:
HAWK 200
   ZH200, the second prototype Hawk 200 single seat fighter seen here while on a test flight from Dunsfold in Spring 1987.

Above:
HARRIER GR.5
   ZD318 being flown by Mike Snelling. First flown on 30 April 1985 it was one of two development batch machines. GR.5 is the RAF version of the Harrier II jointly developed by BAe and McDonnell-Douglas; the first of sixty production GR.5's being delivered to RAF Wittering early in 1987.

Below:
EAP
   Experimental Aircraft Programme technology demonstrator caught by the camera in August 1986. Powered by two RB199 Mk.104D engines it is a precursor of the European Fighter Aircraft: a new breed of supersonic fighter.

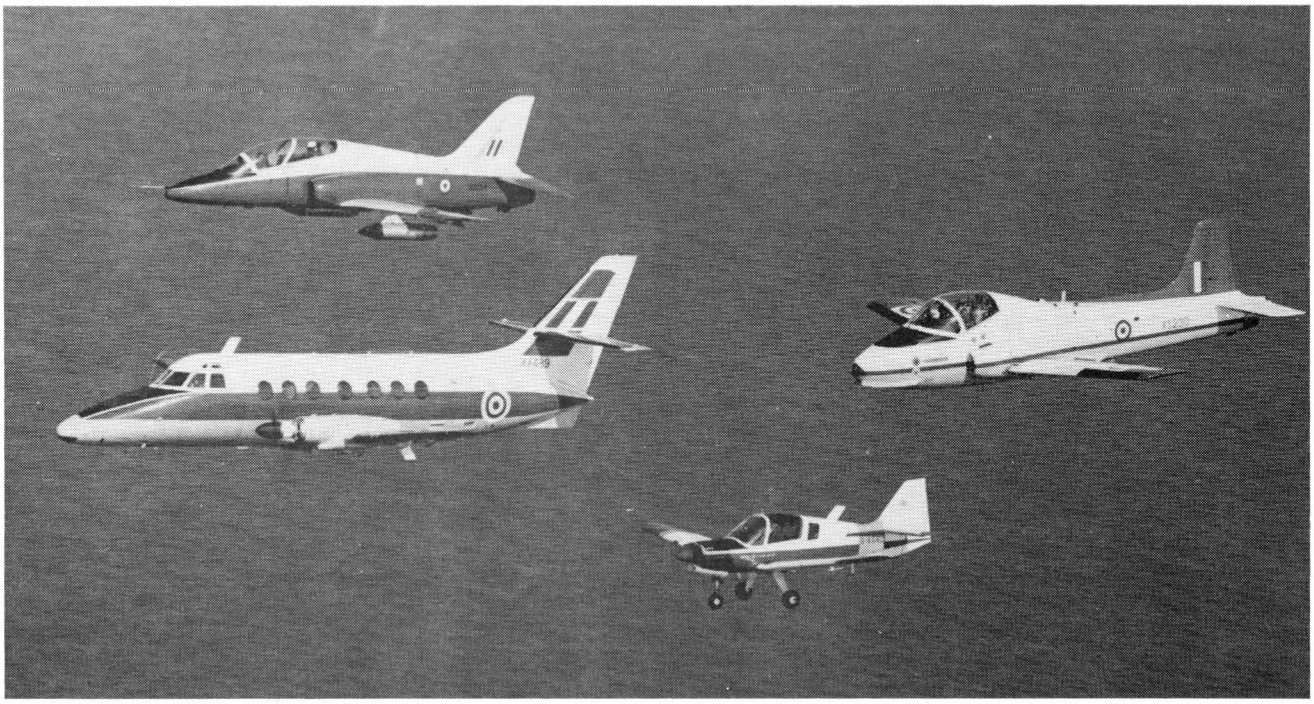

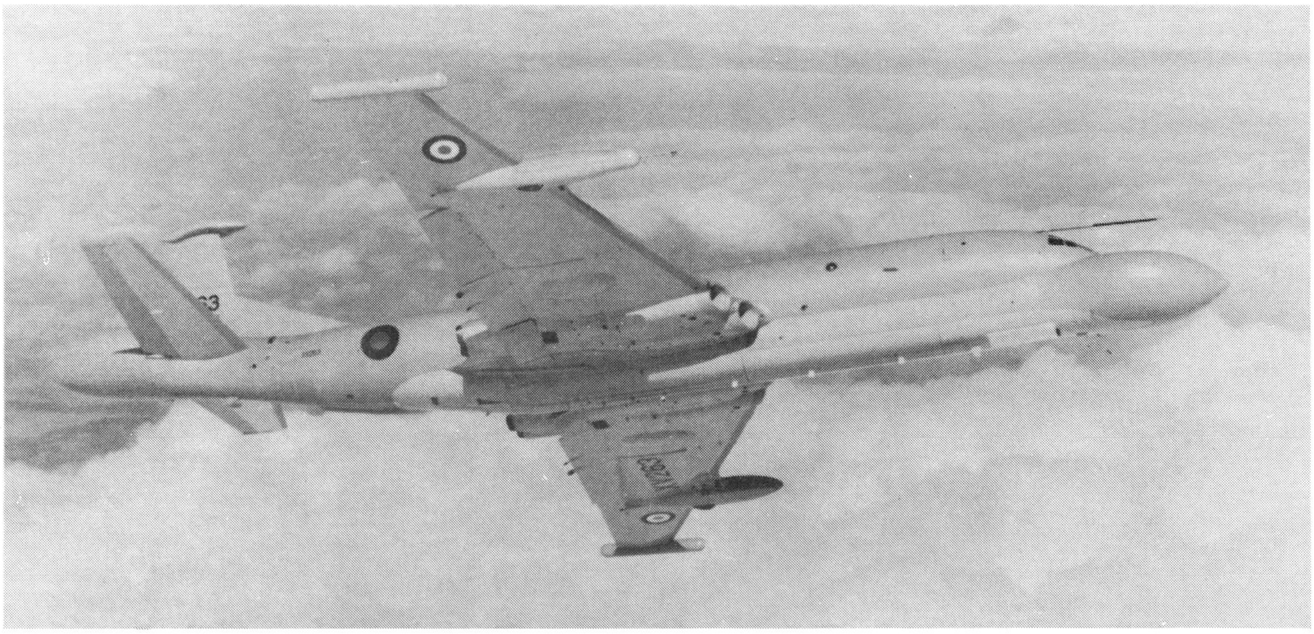

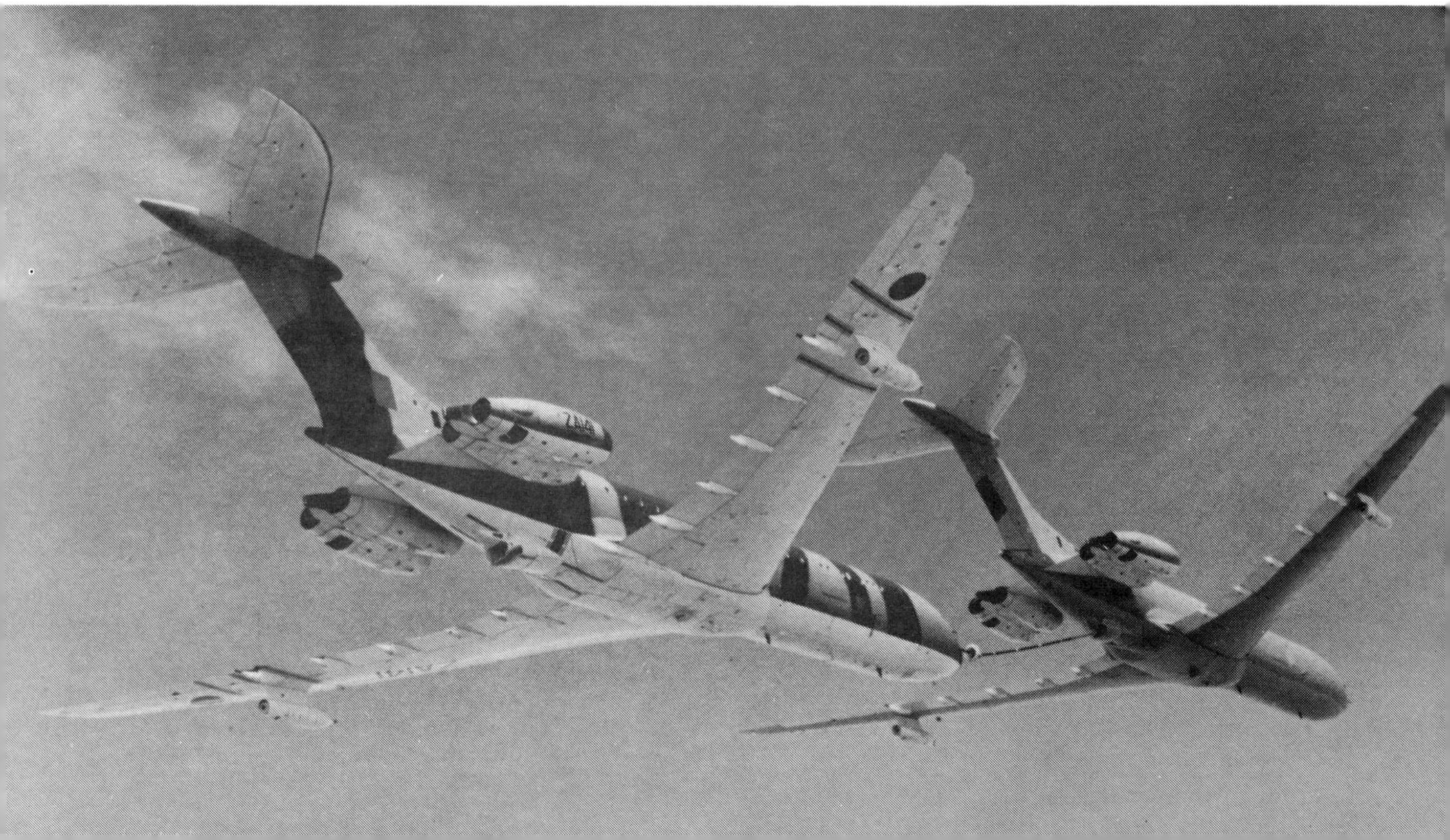

Left Upper:
TRAINERS
Four aircraft from the range of trainers inherited by BAe. Pictured from left to right in 1977 are Jetstream T.1.(XX489), Hawk T.1. (XX154), Bulldog (G-AGAL) and Jet Provost T.4. (XS230).

Left Centre:
NIMROD A.E.W.
XV263 started life as a de Havilland Comet airliner and was rebuilt at Woodford with a secondary fuselage together with nose and tail radomes. Intended as an Airborne Early Warning sentry post, the airframe and engines performed perfectly but the project was cancelled in 1987.

Left Lower:
VC10 AART
ZA141 in camouflage takes up position for refuelling from the ventral hose drum unit of ZA142 which is in the later hemp coloured livery. Conversion of the retired VC10s for RAF use as air-to-air refuelling tankers was Filton's major task following the run down of Concorde production.

Top:
'EUROFIGHTER'
Full scale model of the proposed EFA in pale grey air superiority camouflage. The aircraft is carrying a representative weapon load of AMRAAM, ASRAAM and extended fuel tanks.

Above:
GOSHAWK
Full size replica of BAe/McDonnell Douglas T45A Goshawk being developed to meet the needs of the U.S. Navy for an advanced jet trainer. A descendent of the Hawk, over 300 aircraft will be assembled at Long Beach, California.

Below:
F-111
BAe Filton is responsible for the depot maintenance tasks of all American F-111 bombers based in Europe. Pre-delivery functioning checks are done in this cold soak chamber which simulates conditions at high altitude.

DYNAMICS

Dynamics products receive less public attention than aircraft for several reasons not least of which is the fact that they are often secret in the development stages. Once deployed, they often fly too fast or go deep into space where they cannot readily be seen. Some very intricate parts are never seen at all because they are enclosed in a larger machine. However, the conversion factor of dynamics products, that is the ratio between the value of the raw materials and the finished object, is often far greater and thus more profitable than with aircraft. They can be of immense value in securing a company's future.

When originally formed in 1978, Dynamics Group had two arms; Stevenage-Bristol Division covering what had formerly been the Guided Weapons Division of BAC, and Hatfield-Lostock Division previously Hawker Siddeley Dynamics.

In 1980, Dynamics Group was reorganised into five self accountable business units: ground launched weapons (Stevenage Division), air launched weapons (Hatfield Division) and ship launched weapons (Bristol Division). Stevenage and Filton also had a joint interest in the Space and Communications Division while Lostock was a manufacturing unit to support the weapons divisions. Bracknell Division was added in 1982 on the acquisition by BAe of Sperry Gyroscope in the UK.

Four unit names changed in 1 July 1985 from geographical to product related titles but the site responsibilities remained generally the same for the first ten years.

On 1 January 1988, all weapons and electronics activity was merged into a single Dynamics Division having its H.Q. at Stevenage 'A' site. Space and Communications Division continued as a separate entity at Stevenage 'B' (H.Q.) and Filton.

Above:
VEMS
The Versatile Exercise Mine System provides realistic training and evaluation of mine warfare.

Below:
GYROSCOPES
The Microflex rate-gyro which measures only 22mm in diameter (shown here with its electronic unit) is one of BAe's smaller products. BAe is among Europe's largest gyro manufacturers with an output of about 20,000 units per year.

Right:
BLACK BOXES
The SCR300 flight data recorder and data acquisition unit designed and built at Bracknell is a microprocessor based system for fighter applications and weighs only 10kg.

Below:
TRACKED RAPIER
Mounted on an RCM748 amphibious, armoured vehicle Tracked Rapier is a mobile anti-aircraft system for protecting NATO tanks and troops in the battlefield.

Top:
SEA EAGLE
Long range, air launched, sea skimming, anti ship weapon in service on RAF Buccaneers (as shown) and Royal Navy Sea Harriers.

Above:
ALARM
Air Launched Anti Radar Missile carried by strike and close support aircraft for attacking enemy radars; shown here on Tornado ZA354.

Right Upper:
GIOTTO
Giotto spacecraft being checked out at Filton where it was built. It intercepted Halley's Comet 150 million kilometres from Earth on 13 March 1986 and sent back scientific data and pictures.

Below:
SEA SKUA
A helicopter launched anti ship weapon which has been combat proven on Royal Navy Lynx helicopters.

Bottom:
SEA WOLF
The six missile GWS25 system is operational on Royal Navy Type 22 and Leander-class frigates. A vertically launched VL Sea Wolf (GWS26) has also been developed.

Below:
SEA DART
Powered by a R.R. Bristol Odin ramjet and solid rocket booster, the Sea Dart GWS30 anti-aircraft and anti-ship missile is in service on HMS Bristol, Royal Navy Type 42 destroyers and Invincible-class aircraft carriers.

CIVIL

An agreement signed with Romania in 1978 for the supply of *One-Elevens* led to eventual manufacture under licence in Bucharest; the first ROMBAC *One-Eleven 500* flying for the first time from Hurn in 1982.

1982 also saw the first flight at Filton on 22 June of ZA141, the first of nine VC10 airliners converted to aerial refuelling tankers for the RAF.

The first flight at Hatfield on 3 September 1981 of the BAe 146 feeder-liner marked the launch of the first new British airliner for nearly twenty years. Two versions were offered: the 82-93 passenger 100 Series and a larger 100-109 seat 200 Series which went into production during 1982. Dan Air put the first BAe 146-100 (G-BKHT) into revenue service on 27 May 1983 while on the other side of the Atlantic, Air Wisconsin introduced the first 200 Series exactly one month later. Extension of the model range so far has included the 'stretched' 300 Series and a freighter conversion, 200QT, tailored to the needs of the Australian line *TNT*. To increase production rates from 27 to 40 aircraft per year a second assembly line was established at Woodford in 1987.

On a smaller scale, production of the *Jetstream 31* using Garrett AiResearch turboprop engines received the go ahead early in 1981 and the first production aircraft, G-TALL, was rolled out at Prestwick on 26 January 1982. *Jetstream 31* has proved to be a good export earner with 95 percent of sales made to American customers.

Over twenty years after its maiden flight, the *125* business jet is still going strong with sales approaching seven hundred aircraft. *Viper* turbojets were replaced by Garrett turbofans in the 700 Series which has been the most successful variant so far. The current model, 125-800, first flew on 26 May 1983.

The *ATP* (Advanced Turbo-Prop), one of the latest members of the Company's family of civil aircraft, made its maiden flight in August 1986. Although outwardly similar to the BAe 748, the *ATP* is about 85 percent new and includes composite materials in its airframe. Powered by two Pratt & Whitney of Canada engines driving BAe/Hamilton-Standard 6-bladed airscrews it can carry up to 72 passengers. Launch customer British Midland Airways put the second *ATP* into service in spring 1988.

Above:
125-800
Airframe no.258003 was registered G-BKUW in August 1983 but had first flown with the 'B Conditions' serial G-5-20. It later went to the USA where it became N800BA.

Left Upper:
BAE 748-2B
Company demonstrator G-BGJV in 'house' livery. The -2B offered engineering improvements to give greater fuel efficiency and lower operating costs together with changes to increase its passenger appeal.

Left Lower:
JETSTREAM 31
Company demonstrator G-JSSD, with a BAe logo on the tail fin yet with a Scottish Aviation coat of arms on its engine nacelles, shows the graceful lines of the Jetstream 31 in production at Prestwick. Garrett AiResearch engines are fitted as standard on the 31's.

Below:
BAe 146
The second 146-100 prototype has had many identities. Originally registered G-BAIE in July 1980 it became G-SSHH, as seen here, in May 1981 and G-OPSA in January 1984. During its American sales tour it carried the serial N5828B.

ORGANISATION

To streamline the management of BAe, the number of aircraft divisions was reduced from six to three on 1 March 1984. Until then, the arrangement had been Weybridge-Bristol Division (the old BAC Commercial Aircraft Division including Hurn), Hatfield-Chester Division (the old de Havilland strongholds), Manchester Division (Chadderton and Woodford), Kingston-Brough Division (which included Hamble and Dunsfold), Warton Division (Preston, Warton and Samlesbury) and Scottish Division. The new alignment was Civil Division (Filton, Chester, Hatfield and Prestwick), Warton Division (Preston, Warton and Samlesbury) and Weybridge Division (Weybridge, Brough, Chadderton, Woodford, Kingston, Hamble, Dunsfold and Hurn).

The Groups were disbanded on 1 June 1985 and on 1 July of that year Civil Aircraft Division was formed. The number of aircraft divisions was reduced still further to two by the merging of Weybridge and Warton Divisions into a new Military Aircraft Division on 1 January 1986. On that date Chadderton and Woodford became part of the civil operation.

Rationalisation caused the closure of three small aircraft sites at Holme-On-Spalding-Moor, Bitteswell and Bracebridge Heath in 1982. Hurn closed in 1984 followed by Weybridge in 1987.

Above:
BAE146-300
First of the 120 to 130 passenger 146-300 airliners which first flew in May 1987. The fuselage of the -300 is 10ft longer than in the -200 version.

Below:
ATP
G-MATP, the prototype ATP, pictured over Lincolnshire on 6 August 1986 during its maiden flight. The Woodford-assembled Advanced Turbo Prop airliner bridged the gap between the Jetstream and BAe 146.

FUTURE

The true character of British Aerospace has yet to emerge and its story will be a subject for future historians. It has, however, been appropriate in reviewing the history of British Aerospace to chronicle, briefly, the progress in the first ten years of its existence and to present a montage of the BAe products of the 1980s and beyond.

Above:
AIRBUS A300
Lufthansa A300-600 with pre-delivery registration F-WWAL in 1987. The stretched A300 can carry 267 passengers. *Airbus Industrie.*

AIRBUS

British involvement in the Airbus project dates back to the mid-1960s when Hawker Siddeley first exchanged ideas with other European manufacturers. In 1967 the British, French and German governments signed a tripartite Memorandum of Understanding which initiated the design of the A300.

The British Government announced in March 1969 that it was withdrawing its support for the project because there were no buyers for either the original 300 seat proposal or the revised 250 seat airliner.

On 29 May 1969 the Airbus programme was launched jointly by France and Germany but Hawker Siddeley designed and built the wings for the smaller A300B as a private venture until official interest was revived.

Britain became a full member of Airbus Industrie on 1 January 1979 when the European consortium was reorganised; the British share representing 20 percent compared to 37.9 percent each for France and Germany and 4.2 percent for Spain.

With an established range of three airliners and others to follow, Airbus is seriously challenging the Americans for this potentially lucrative market.

Right Centre:
A310 WING
The first A310 wing box to be built at 'Chester' pictured on 7 April 1981. On completion, it became part of Swissair's A310, registered HB-IPE.

Right:
AIRBUS A310
Swissair A310 with French registration F-WZLH. A310 the smaller short-medium range partner of the A300, seats 225 people in a wide bodied fuselage and commenced flight testing on 3 April 1982. *Airbus Industrie.*

Above:
Airbus wing assembly at Broughton.

Right:
AIRBUS A320
The third flight test A320 (F-WWDB) in the colours of the principal French domestic airline, Air Inter, in June 1987.

Below:
WING CENTRE
A320 wings being equipped in the new wing centre at Filton in 1987 prior to airlift by Guppy to the final assembly line in Toulouse.

Below Left:
AIRBUS A330
Twin engined medium range airliner being developed to carry 328 passengers.

Below Right:
AIRBUS A340
The first four-engined Airbus being developed in two versions to carry up to 295 passengers on long haul routes.

Above:
SABA
Small, Agile Battlefield Aircraft proposed for the 1990s.

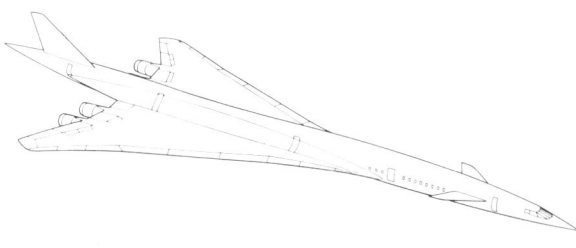

Above:
AST
An idea of what an advanced supersonic transport might look like in the next century. AST would be about twice the size of Concorde.

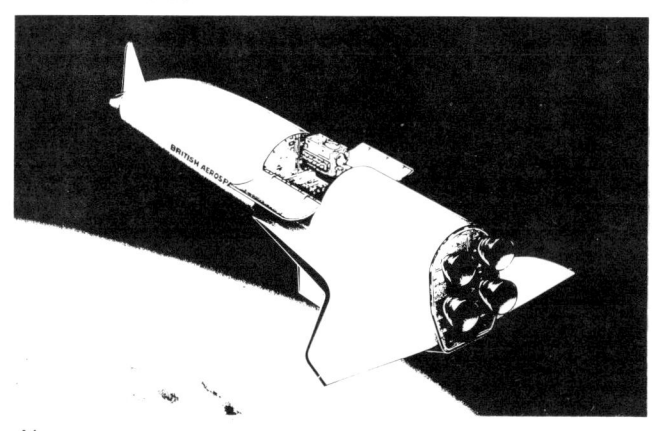

Above:
HOTOL
For economical satellite launching purposes, BAe proposes an unmanned horizontal take off and landing aerospaceplane powered by a Rolls Royce RB545 hybrid rocket engine. First flight could take place before the end of this century.

Right Upper:
146 DEVELOPMENT
Artist's impression of a BAe146 having two open-rotor engines shows how the high wing of the 146 allows almost any powerplant with adequate power to be used.

Right Centre:
BAe.125
An artist's impression of what could be the ultimate development of the '125' series of business jets.

Right Above:
JETSTREAM 41
Artist's impression of a development of the Jetstream 31 small airliner with seating capacity increased from 19 to 27 passengers.

Right Lower:
'MINI CONCORDE'
Artist's conception of a supersonic business and executive jet capable of carrying twelve passengers at speeds up to Mach 1.85 for 3,800 miles.

Into The 1990s

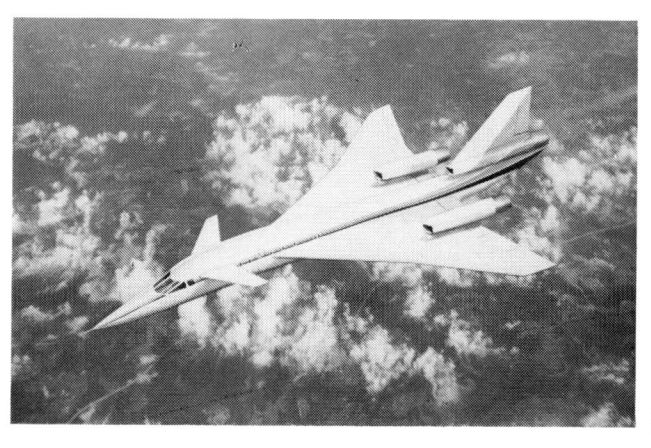

Factory Locations in the U.K.

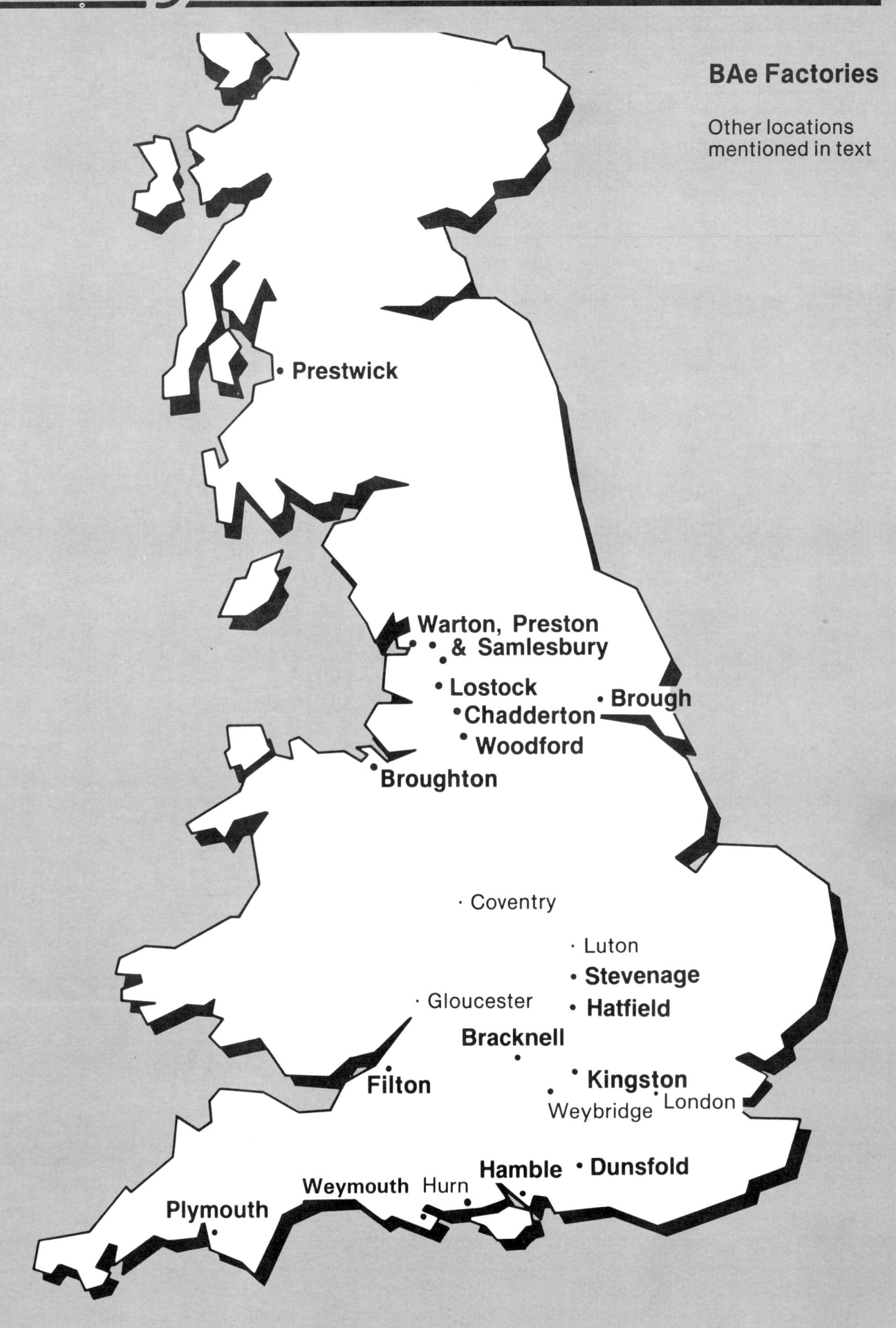

BAe Factories

Other locations mentioned in text

- Prestwick
- **Warton, Preston & Samlesbury**
- **Lostock**
- **Chadderton**
- **Brough**
- **Woodford**
- **Broughton**
- Coventry
- Luton
- **Stevenage**
- Gloucester
- **Hatfield**
- **Bracknell**
- **Filton**
- **Kingston**
- Weybridge
- London
- **Hamble**
- **Dunsfold**
- Weymouth
- Hurn
- Plymouth